INTEGRATING N
INTO MENTAL
CARE

Mental health counselors, psychologists, social workers, and psychiatrists realize that nutrition may be a factor in their clients' mental health, but a lack of nutritional science background and resources makes it difficult for them to incorporate nutrition into the care they provide. Likewise, registered dietitian nutritionists new to the field of mental health care (whether in a facility or in private practice) may feel the need for succinct resources geared to this area of nutritional care. *Integrating Nutrition into Mental Health Care* illuminates the intersection between nutrition and mental health, bridging the gap for professionals in both fields.

It presents resources in areas such as caffeine intake, family history of a genetically transmitted nutrition-related condition, interpretation of laboratory nutritional assessment, and safe upper limits of supplements, as well as additional nutrition factors, helping practitioners easily incorporate selected nutritional aspects into the mental health care of clients. The book includes sample forms for office use and instructions to interpret client information. They can be easily downloaded and printed from the Routledge book page. Additional forms available help in facilitating referral to a registered dietitian nutritionist (RDN) for a deeper look at individual patient's nutritional issues.

To offer further information on nutrition and mental health, the text features a series of short fictional, but real-life, nutrition stories. These stories provide an enjoyable format in which to train, or test, the reader's knowledge.

INTEGRATING NUTRITION INTO MENTAL HEALTH CARE

Ruth Leyse Wallace, PhD

CRC Press
Taylor & Francis Group
Boca Raton London New York

CRC Press is an imprint of the
Taylor & Francis Group, an **informa** business

Designed cover image: Shutterstock

First edition published 2025
by CRC Press
2385 NW Executive Center Drive, Suite 320, Boca Raton FL 33431

and by CRC Press
4 Park Square, Milton Park, Abingdon, Oxon, OX14 4RN

CRC Press is an imprint of Taylor & Francis Group, LLC

ISBN: 978-1-032-88694-7 (hbk)
ISBN: 978-1-032-84272-1 (pbk)
ISBN: 978-1-003-53915-5 (ebk)

DOI: 10.1201/9781003539155

Typeset in Caslon
by SPi Technologies India Pvt Ltd (Straive)

Access the Support Materials: https://www.routledge.com/9781032842721

Contents

V

But Officer, I Would Never
Choiceville, USA
There's a Method to Her Madness
The Camp Conundrum
Step Right Up – Play Life Today
Too Much or Not Enough?
The Nonanonymous Support Group

List of Tables
List of Forms and Assessments
Dietary Guidelines for Americans 2020–2025 Executive
Summary (Excerpt)
Common Abbreviations
Website Resources
References
    References for Chapter I – Why Integrate Nutrition Care
    and Mental Health Care?
    Additional Reading
    References for Table 2: Reports and Observations That
    May Be Related to Nutritional Risk
    References for Laboratory Assessment of Nutritional
    Status
    References for the Nutrition-Focused Physical
    Examination
    Selected References for the Gut, Microbiome, and Brain
    Selected References for Table 8: Nutrients, the Brain,
    CNS, and Mental Status

About the Author

Ruth Leyse Wallace, PhD, received her BS degree from the University of California at Davis, earned her MS degree while completing her dietetic internship at the University of Kansas Medical Center in Kansas City, and was awarded her PhD from the University of Arizona in Tucson. Her dissertation proposed and validated a Metaparadigm of Clinical Dietetics, an organizing structure for the body of knowledge used by clinical dietetics.

She practiced clinical dietetics in the areas of mental health, eating disorders, substance abuse, and general psychiatry at Osawatomie State Hospital in Osawatomie, Kansas; at The Menninger Foundation in Topeka, Kansas; at Sharp Mesa Vista Hospital in San Diego, California; and at Hospital Corporation of America (HCA) Hospital in Plano, Texas. While in Topeka, in the early 1980s, she established one of the first private practices for nutrition counseling in the state. During this time, she was commissioned by Fleming Food company (part of IGA – Independent Grocers of America), to create one of the first sets of consumer nutrition education materials to be exhibited in grocery stores. Dr. Leyse Wallace served as an adjunct faculty member at Pima County College in Tucson and Mesa Community College in San Diego.

Dr. Leyse Wallace has published three books: *Nutrition and Mental Health* (CRC Press, Taylor and Francis, Publishers, Boca Raton, Florida), *Linking Nutrition to Mental Health: A Scientific Exploration* (iUniverse, Inc., Lincoln, Nevada), and *The Metaparadigm of Clinical Dietetics: Derivation and Applications.* (also with iUniverse, Inc.), as well as chapters in two nutrition textbooks: *Integrative and Functional Medical Nutrition Therapy – Principles and Practices* (Chapter 30: edited by Diana Noland, RDN, Jeanne Drisko, and Leigh Wagner Maryland, 2020) and in *Krause's Food and the Nutrition Care Process* (Chapter 41: "MNT in Psychiatric and Cognitive Disorders" by Teitelbaum, Jacob, Alan Weiss, Geri Brewster, and Ruth Leyse Wallace. Kathy McMahon, editor).

A 50-year member of the American Dietetic Association (ADA), now the Academy of Nutrition and Dietetics, she has been an active contributor to the Behavioral Health Nutrition (BHN) dietetic practice group in the Academy serving as Mental Health Resource Professional on the Executive Committee and as co-author of the 2018 revision of the Standards of Practice and Standards of Professional Performance for the BHN dietetic practice group. Dr. Leyse Wallace has contributed articles to the group newsletter and presented educational webinars for BHN. In 2010, she was presented the BHN Excellence in Practice award.

Among others, Dr. Leyse Wallace has presented at meetings of the National Alliance on Mental Illness (2012), the Department of Integrative Medicine at University of Kansas Medical Center (2013), and the American Society of Parenteral and Enteral Nutrition (2003) and a Symposium on the Nutrition Screening Initiative in Washington, DC (1997).

In 2017, Dr. Leyse Wallace was the recipient of the Albert Nelson Marquis Lifetime Achievement Award by the Marquis Who's Who Publications Board.

Now living in Chandler, Arizona, Dr. Leyse Wallace continues to enjoy writing and corresponding with colleagues, friends, and family.

I

WHY

Why Integrate Nutrition Care and Mental Health Care?

Nutrition and Mental Health Are on the Public Mind

A survey reported by *Parade* magazine, found that more than twice as many Americans (62%) fear losing their mental capacity as fear losing physical ability (29%) as they age; this justifiably concerned, it seems. In 2015, 43 million (17%) adults in the United States had mental disorders such as anxiety, autism, bipolar disorder, borderline personality disorder, and psychosis. Over eight million citizens had dual diagnoses (such as depression and an addiction).

The concern is obviously in the public mainstream when nutrition themes appear on TV shows. For example, on one show, a suspect was psychotic, which was thought to be due to vitamin B-1 deficiency. On another show, a victim was killed because of the interaction between her depression medication (monoamine oxidase inhibitor) and fermented balsamic vinegar. In another television drama, a child with scurvy is brought to the emergency room for diagnosis. Cartoonists in newspapers and magazines include themes of diet, weight, snacking, and eating disorders – another indication that nutrition and mental health are on the minds of the public.

History and Current Observations

In 1944, after World War II, as part of the program to aid in the restoration of Europe, Ancel Keys, at the University of Minnesota, conducted an experiment on 36 conscientious objectors to study the physical and psychological effects of human semi-starvation and refeeding. The study increased understanding that starvation dramatically alters personality and that nutrition directly and predictably affects the mind as well as the body. It changed scientific attitudes regarding the fact that diet alone could affect areas previously considered fixed: blood pressure, cholesterol, and heart rate.[1,2] Roy-Byrne, Gorelick, and Marder, researchers at the UCLA School of Medicine, report on the hospitalization of a young male, linking symptoms of vitamin C deficiency to his mental state and schizotypal personality disorder. They conclude, "Psychopathology and nutritional status are interdependent. A patient's mental state may contribute to unusual eating habits, leading to nutritional deficiencies, which may in turn lead to biochemical changes, which could exacerbate psychopathology already present. Nutritional status, eating habits and dietary intake are an important consideration in evaluating psychiatric patients."[3] Although the determinants of mental health are complex, the emerging and compelling evidence for nutrition as a crucial factor in the high prevalence and incidence of mental disorders suggests that diet is as important to psychiatry as it is to cardiology, endocrinology, and gastroenterology."[4] People feel worse, can't think as clearly, function poorly, and often have less mental energy when their nutritional needs are not met. Poor nutrition can affect intellect, personality, mood, and cognition.

> Vitamin deficiency diseases usually involve both the central and peripheral nervous system, unlike many other diseases.
>
> **Nutritional and Metabolic Diseases of the Nervous System.** in *Harrison's Principles of Internal Medicine* (1994, p. 2329–2333).

The Quality of the American Diet Today

The 2015 Dietary Guidelines for Americans[5] report that large percentages of citizens do not meet the recommendations for food and nutrient intake. Nutrients considered in "shortfall" status include vitamins A, C, D, E, folate, calcium, fiber, potassium, and magnesium. For adolescent and premenopausal females, iron is also not adequate. Of these nutrients, calcium, vitamin D, fiber, and potassium are also classified as nutrients of public health concern because their under-consumption has been linked in the scientific literature to adverse health outcomes. The majority of the U.S. population has low intakes of key food groups that are important sources of the shortfall nutrients, including vegetables, fruits, whole grains, and dairy foods.[6]

Mental Energy: the ability to perform mental tasks, the intensity of feelings of energy/fatigue, and the motivation to accomplish mental and physical tasks. Core dimensions include cognition, mood of energy, and motivation. Variables that can influence mental energy include genetics, nutrition, pain, sleep, and others.

ILSI. Mental Energy: Defining the Science. Inter-national Life Sciences Institute Workshop. Washington, DC. Special Supplement to *Nutrition Reviews*. July 2006; 64 (7)(Part II)(S1–S16).

U.S. Dietary 2010 Guidelines report that serum vitamin C deficiency, was found in 7.1 percent of the population of the United States. Based on the 2019 population of 328.2 million that amounts to 23.3 million people in the United States being deficient in vitamin C. Personality changes such as hypochondria, depression, and fatigue occur at 1.21–1.17 mg/100 ml, earlier than the physical signs and symptoms of scurvy, which occur at 0.67–0.14 mg/100 ml.[7]

A thought to remember: "Partial and acute vitamin deprivation suggest that the earliest impairments occur in measures of mood rather than mental performance."
Haller, Jurg. "Biokinetic Parameters of Vitamins A, B-1, B-2, B-6, E, K and Carotene in Humans."

Nutritional Neuroscience, 2005. Harris R. Lieberman, Robin B. Kanarek and Chandron Prasad, editors. CRC Taylor and Francis, New York. p. 229.

It is reported that 10%–20% of people in their late 60s or older have some degree of B-12 deficiency and that 800,000 elderly people in the

United States have undiagnosed and untreated pernicious anemia.[8] This deficiency was also reported in the case study of a 6-year old adolescent discovered during admission to a psychiatric treatment facility.[9]

If researched, it wouldn't be surprising for patients of today's mental health care providers to have nutrition questions or health issues related to nutrition.

Mental Health Care Providers Weigh In

A dissertation survey of mental health professionals (N = 86 respondents) by Kristin Sliter, EdD, indicated interest in nutritional issues, positive beliefs, and attitudes (98%) regarding healthy dietary practices, and indicated belief that mental health is connected to nutrition (97%).[10]

Using open-ended questions, the survey of mental health care providers (N = 86) reported eight major findings:

1. An overwhelming majority (98%) of responding mental health professionals reported a favorable attitude about nutrition and healthy dietary practices.
2. Ninety-seven percent of respondents believe nutrition is connected to mental health.
3. An overwhelming majority of responding mental health professionals believe they have a role in promoting healthy dietary practices.
4. Sixty-nine percent of respondents did not believe they have the necessary knowledge to discuss nutrition with their clients.
5. Approximately half of mental health professionals who completed the survey conduct dietary assessments.
6. An overwhelming majority of responding mental health professionals believe in nutritional counseling as an adjunctive intervention.
7. Over half (62%) of respondents implement some form of nutrition-related interventions based on primarily informal assessments.
8. Many mental health professionals acknowledged barriers to discussing nutrition-related topics.

These findings suggest that health care professionals are convinced that there is a relationship between nutrition and mental health but need some tools to support their interest in acquiring nutrition information from their clients and guidelines on how to interpret and use what they discover. Nutrition is only one influence, but it is important to determine that factors such as nutrient deficiency, excess, or imbalance are *supporting* mental health instead of *interfering* with function, mood, and/or cognition.

It's a Two-Way Street

The quality of an individual's diet and nutritional status influences their mental status; and it is bidirectional. Mental status can influence the following:

- Desire to eat
- Ability to attend to obtaining and preparing food
- Energy needs
- Consistent intake of recommended medications and supplements
- Need for medications that affect nutrient metabolism
- Willingness to accept foods that don't meet a personal criterion—for example, fear of being poisoned, being surreptitiously given medication
- Fear of being unable to swallow

Since nutrient deficiencies and excesses result in changes in mental status, and changes in mental status influence nutritional intake, and a less-than-ideal diet isn't uncommon, doesn't it make sense that assessment and treatment for mental health issues should include assessment for nutritional issues?

Research in Nutrition and Mental Health

Research in mental health and nutrition can follow many avenues from individuals to populations. Ideas for research in private practice, institutions, community mental health services and other settings include the following:

1. Homeless populations are likely to have nondependable food supplies; they are also often found to suffer from mental illness and other mental and physical health problems. With these coexisting conditions, and considering the mental effects of poor nutrition, what could be observed, documented, and addressed if professionals trained in conducting nutrition focused physical examinations (NFPEs) were to conduct them at health fairs, clinics, or other community venues?[11]

2. Acute hospitals measure length of stay (LOS) and readmission rates and found these measures are correlated with nutritional status. What would we find if these were measured for populations in psychiatric hospitals, outpatient clinics, addiction treatment centers, etc.? Would those who were well-nourished need less frequent readmissions and/or follow-up?

3. What proportion of patients admitted for psychiatric treatment have physical signs/symptoms of poor nutritional status?

4. Which nutrient/nutritional assessment laboratory tests are obtained routinely at psychiatric facilities? Which are available to professionals, mental health practitioners, or registered dietitian nutritionists (RDN) in private practice?

5. Research in psychology and psychiatry often includes a comparison of two groups. To ensure a lack of selection bias, participant's nutritional status should be determined so it is known to not confound results. Comparisons of nutritional equality are seldom reported in research results.

6. Many nutrients play a part in the structure and function of the body: our DNA, organs, hormones, the brain, and bodily functions. Table 8 describes some of these functions and how specific nutrients influence our brain and mental status. Nutrition-related research often tells us "how" nutrients influence our biochemistry and other functions, but the specific mechanisms are not known in some cases. Scientific evidence may or may not support some popular claims for foods, diets, and supplements.

Kaplan[1] proposed four models to possibly explain how nutrients may ameliorate mental symptoms related to (1) expressions of inborn errors of metabolism, (2) manifestations of deficient methylation reactions, (3) alterations of gene expression by nutrient deficiency, and (4) long latency deficiency diseases.

An Overview – How Nutrients Affect Mental Health

1. Nutrients and other elements may support or interfere with normal development and maintenance of the brain and central nervous system of the fetus, growing child, or adult. (Essential fatty acids are essential to myelin sheath formation.
   Altered metabolism and excretion of amino acids or carbohydrates, such as phenylalanine or galactose, may result in accumulation of levels toxic to the brain.)
2. Nutrients may serve as precursors to the manufacture of neurotransmitters. They may contribute skeletons of a molecule or a required component of the neurotransmitter molecule. (Folic acid is a source of methyl groups.)
3. Nutrients are needed to supply the brain with an energy source and the ability to use the energy (Carbohydrate is needed for glucose, which is the brain's primary energy source.)
4. Nutrients may influence genetic transcription. (Nutrients influence metabolic signals for stimulating or failing to signal for transcription of genes.)
5. Nutrients may have pharmacologic functions at doses higher than nutriologic requirements – either to accommodate altered genetic transcription or as toxic elements. (Nutrients may function as pro-oxidants, as well as antioxidants.)
6. Nutrients and food contribute to mood, sense of well-being, and psychological function, perhaps related to changes in nutritional status. (Mood may change as a result of food restriction, dieting, and starvation.)
7. There may be changes in the entrance and/or exit of nutrients through receptors on cell surfaces related to thoughts and emotions. (Stress-related dysfunction of cells and systems may affect nutrient use.)

Read the included "nutrition detection stories" and test your inner nutrition detective.

Note

1 Kaplan, BJ, SG Crawford, CJ Field, and JS Simpson. "Vitamins, Minerals, and Mood." *Psychol Bull*. 2007; 133(5):747–760.

II

HOW TO INTEGRATE NUTRITION CARE AND MENTAL HEALTH CARE?

Components of Nutrition Care

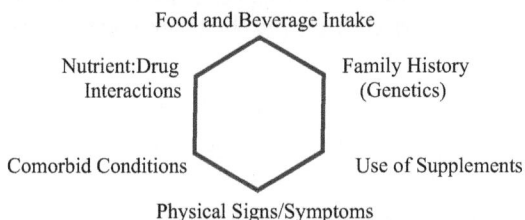

Food and Beverage Intake

Nutrient:Drug
Interactions

Family History
(Genetics)

Comorbid Conditions

Use of Supplements

Physical Signs/Symptoms

Table 1 Reports and Observations That May Be Related to Nutritional Risk

SIGNS/SYMPTOMS OBSERVED OR REPORTED	DESCRIPTION	RULE OUT INADEQUACY OR EXCESS
General fatigue	Energy metabolism; availability of ATP for cellular energy	
	Carbohydrate, glucose to ATP	Thiamin (B-1)
	Fat, fatty acids to ATP	Riboflavin (B-2)
Fatigue, headache, decrease in memory and concentration		Mercury toxicity
Personality changes	Personality changes occur earlier in vitamin C deficiency than the physical signs of scurvy	Vitamin C
Depression		Omega-3 fatty acids, zinc, folic acid, B-12

(Continued)

DOI: 10.1201/9781003539155-2

Table 1   (Continued) Reports and Observations That May Be Related to Nutritional Risk

SIGNS/SYMPTOMS OBSERVED OR REPORTED	DESCRIPTION	RULE OUT INADEQUACY OR EXCESS
Atypical depression, a common form of major depression, is characterized by a distinct combination of symptoms including mood swings, carbohydrate cravings, weight gain, rejection sensitivity, and lethargy		Chromium
Ecchymoses; easily bruised	Under the skin bleeding; fragility of red blood cells is among the causes of bruisability	Vitamin C deficiency; Scurvy
Perifollicular petechiae	Red areas around hair follicles	Vitamin C
High alcohol intake		Thiamin, beriberi, niacin (B-3) pellagra, pyridoxine (B-6)
Alcohol intake withdrawal		Niacin, thiamin
Seizures during alcohol withdrawal		High copper, low zinc
Confusion, apathy, inattentiveness, disorientation to time and place, inability to grasp the meaning of the immediate situation, and irrational inconsistent remarks	Wernicke-Korsakoff syndrome; alcohol intake The biosynthesis of acetylcholine and GABA (gamma-aminobutyric acid)	Inadequate thiamin
Foot drop, abnormal gait, ataxia, wrist drop (following bariatric surgery)	Effects nerve and muscle function; flow of electrolytes in and out of cells Wrist drop: Flexion: N = 60–80 degrees (tilting toward the palm) Dorsiflexion/extension: N = 60–75 degrees (tilting toward the back of the hand)	Inadequate thiamin
Persistent vomiting	Following bariatric surgery and during pregnancy	Thiamin
Paleness, tiredness; cold hands and feet, shortness of breath	Macrocytic anemia; megaloblastic anemia Microcytic anemia	Folate/folic acid and/ or B-12 Iron
Underweight, missing meals due to lack of money		General malnutrition

(Continued)

Table 1 (Continued) Reports and Observations That May Be Related to Nutritional Risk

SIGNS/SYMPTOMS OBSERVED OR REPORTED	DESCRIPTION	RULE OUT INADEQUACY OR EXCESS
High sugar intake, high intake of refined and unfortified carbohydrates		High-calorie malnutrition
(1) Autonomic (palpitations, sweating, shaking, and hunger) (2) Neuroglycopenic (confusion, drowsiness, inability to concentrate, speech difficulty, and blurred vision) (3) Nonspecific (nausea and headache)	Hypoglycemia: blood glucose level below 70 mg/dl.	Inadequate glucose/ carbohydrate
Fatigue, food cravings, anxiety, compulsive eating, nervousness, confusion with extremely low or excessive glucose levels		
Hyperglycemia: Individuals may be unable to concentrate, unable to pay attention, may be unusually moody and irritable	Hyperglycemia: Blood glucose level above 250 mg/dl has been observed to change mental status May be classified as being (1) alert, (2) drowsy, (3) in a stupor, or (4) in a coma	Glucose exceeds capacity of insulin to metabolize carbohydrate
Redness, irritation surrounding orifices (mouth, nostrils, eyes, ears)	Decreased sense of taste	Inadequate zinc
Diarrhea		Excess zinc
Irritability, nervousness and restlessness	Very high intake of fiber inhibit iron and zinc absorption (cereals, legumes and nuts)	Zinc
Severe neurological symptoms, gastrointestinal and acute respiratory distress syndrome, myocardial infarction, hair loss, muscle tenderness		Selenium toxicity
Memory loss, hyperactivity, aggressiveness, and defensiveness, and may occur at relatively low doses of selenium		Selenium toxicity
Dementia (one cause) Accumulation of copper, copper rings visible in cornea		Wilson's disease

See Appendix for References for Table 1.

Assessment of Food and Beverage Intake

Form 1. Nutrition History: Food, Caffeine, Alcohol, Fluid, Weight p. 1 of 2

Name_____Date_____

Nutrients and nutritional status may affect a person's mental health. Please share the following general information about your food intake and eating habits to help assess whether or not you might benefit from further assessment:

(Check, Circle, or Write in specific foods)

I. **What is the first time you eat during a typical weekday?** _____

On weekends? _____

What do you commonly eat/drink at that time?

*Coffee/Tea Juice Fruit Milk Toast/Muffin/Bagel/Biscuit Cereal Waffle

* Pancake Egg/Ham/Sausage/Cheese Bacon/Butter/Margarine/ Peanut Butter

* Yogurt Smoothie Protein Shake Popcorn Other _____

II. **Do you eat a meal or snack around mid-day (11AM–2 PM?) Yes / No**

What do you commonly eat/drink at that time?

*Milk / Shake or Malt / Juice/ Coffee / Tea / Soft Drink / Beer/ Wine / Water

*Hot or cold sandwich Chips Fries Soup Salad Pickles

*Meat / Poultry / Fish / Bean / Cheese Main Dish / Pizza

*Potato Rice Pasta Bread / Tortilla / Roll / Crackers / Pita / Bagel

*Vegetables – Raw / Cooked / Juice Fruit – Whole / Juice

*Dessert – Ice Cream / Cookies / Cake / Pie / Candy / Pudding / Jello /

Other_____

III. Do you eat a meal or snack between 3:00 and 8:00 PM? Yes / No

What do you commonly eat/drink at that time?

*Milk / Shake or Malt / Juice/ Coffee/ Tea / Soft Drink / Beer/ Wine / Water

*Hot or Cold Sandwich Chips Fries Soup Salad

*Beef / Pork / Poultry / Fish / Bean Main Dish / Pizza

*Potato Rice Pasta Bread / Tortilla / Roll / Crackers / Pita / Bagel

*Vegetables – Raw / Cooked / Juice Fruit – Whole / Juice

*Dessert – Ice Cream / Cookies / Cake / Pie / Candy / Pudding / Jello /

*Other_____

Nutrition History: Food, Caffeine, Alcohol, Fluid, Weight p. 2 of 2

Name_____Date_____

IV. Do you eat a meal or snack in the mid-to-late evening? Yes / No Meal / Snack

What do you commonly eat/drink at that time? (Circle or name specific other foods)

*Chips Sweets / Deserts / Candy / Leftovers / Sandwiches / Cheese / Crackers

*Nuts / Beverages: Coffee / Tea / Beer / Wine / Soft Drink / Milk or Cocoa / Water

Other _____

* I do / don't eat on a regular schedule.

* I do / don't eat very differently on weekdays than on weekends: What changes? Different time? Different foods? Different place?

Other _____

* Some foods that I eat often but that aren't on the lists above:

_____

* Are there any foods you never eat because of allergies, because you dislike them, or any other reason? Yes / No. Which foods?

_____

* Yes / No I often / sometimes have low calorie / low fat or other "diet" foods/beverages such as _____

_____

* Would you say you eat small, medium, or large servings?

My height is ____. My weight is _____lbs.

   I'm OK with this weight. Yes No

   I recently gained/lost _____lbs. I would like to weigh _____.

* **Do you have any additional comments?**

_____

_____

_____

Thank you

Form 2. Family History and Allergies

Do you have any food allergies? No Yes _____

If yes, What foods are you allergic to? _____

Family History: Do you or a relative have any of the following conditions? (Check any that apply)

CONDITION	ME	A RELATIVE? RELATIONSHIP (PARENT, COUSIN, ETC.)
Anorexia Nervosa or Bulimia or Other Eating Disorder?		
Alcohol Intolerance or Abuse?		
Celiac Disease or Wheat Intolerance		
Irritable Bowel Syndrome or Other Intestinal Disorder		
Heart Disease		
Kidney Disease		
Hemochromatosis (Iron Accumulation in the Blood)		
Thyroid Disorder		
Goiter (Enlargement of Thyroid Gland)		
Phenylketonuria (PKU)		
Galactosemia		

1A Assessment of Food Intake and Eating Habits

USUAL HABITS	CRITERIA
_____ #Meals _____ # Snacks eaten /day	Less than 3/day is not likely to be nutritionally adequate Calorie intakes below 1,500/day are unlikely to meet nutrient requirements of adults
Eats all food groups: Meat or Substitutes, Vegetables, Fruits, Grains, Dairy, Fats	___ Omission of _____ ___ " _____ (see below)
Chooses a variety of foods; few repetitive choices	Variety increases inclusion of more nutrients
2 or fewer servings of sweets daily	< 10% of calories is recommended: (example: 200 calories is 10% of a 2,000 calorie/day intake)

1 (F)–2 (M) servings of alcohol/day	A binge is commonly defined as > 5 drinks for males, > 4 drinks for females per drinking episode. Alcohol increases needs and decreases intake of multiple vitamins and minerals
Food Group	**Major nutrients that are contributed by many foods in this group** **May be deficient if food group is omitted or eaten in less than the recommended quantities**
Dairy (Milk, Yogurt, Cheese, Ice cream)	Calcium, riboflavin, protein, carbohydrate (minimal carbohydrate in cheese) Potassium varies in different dairy foods Vitamins D and A (in milk, if fortified)
Beef, Pork, Chicken, Fish, Eggs	Protein, iron, niacin, fat, zinc
Starches	Thiamin, riboflavin, niacin, folic acid, and iron present in enriched carbohydrates Zinc and magnesium present in whole grains but not refined Folic acid and iron provided by peas and beans, legume carbohydrates
Vegetables	Vitamins C and A, folic acid, carbohydrate (in some vegetables), potassium, magnesium (especially in green vegetables)
Fruits	Carbohydrate, Vitamin C, Vitamin A, Potassium
Fats and Oils	Omega-3 and Omega-6 fatty acids, vitamin E

See also

(a) Table 4. Potentially Toxic Minerals/Elements Found in Food,
(b) Table 7. Food Sources of Nutrients, and
(c) Dietary Guidelines for Americans 2020–2025 – Executive Summary

1B Assessment of Alcohol Intake

A "drink" is defined as the amount of the beverage that contains 1.5 oz of pure ethanol. The liver can detoxify 1.5 oz of ethanol per hour.

A standard drink is: One 12-oz can or bottle of beer
5 oz of wine
1.5 oz of 80-proof distilled spirits
Moderate = 2 drinks/day for men; 1 drink/day for women
(The 2015–2020 edition of the Dietary Guidelines for Americans allows two drinks per day for men and was not changed for the 2020–2025 Dietary Guidelines.)

https://www.medscape.com/viewarticle/943344?nlid=
138886_3901&src=wnl_newsalrt_201229_MSCPEDIT&
uac=39259FT&impID=2787555&faf=1

Binge = > 5 drinks/per 2-hour episode for males
> 4 drinks per 2-hour episode for females

Reported drinks of alcohol/day _____

Assessment tools include the AUDIT (10 questions), AUDIT-C (3 questions)

See also

www.integration.samhsa.gov/clinical-practice/screening-tools, as well as the single-question method recommended by NIAAA (National Institute on Alcohol Abuse and Alcoholism): "How many times in the past year have you had 5 or more [for men] or 4 or more [for women, and all adults older than 65 years] drinks in one day?"

Further details about the single-question screening method, as well as resources on primary care–feasible behavioral interventions, are available from the NIAAA at (http://pubs.niaaa.nih.gov/publications/Practitioner/CliniciansGuide2005/guide.pdf).

1C Assessment of Caffeine Intake

Less than 400 mg/day = moderate

_____ C. Coffee (100 mg/5 oz cup)	X	5oz cups =	_____
_____ C. Glasses of tea (8–10 oz each)	X	80 mg =	_____
_____ 12 oz cans cola	X	40 mg/can=	_____
_____ 12 oz can Mountain Dew	X	55 mg/can =	_____
_____ 1 can Red Bull drink	X	80 mg/can =	_____
_____ Cups cocoa (5–8 oz each)	X	25 mg/cup =	_____
_____ 1 oz semisweet chocolate	X	13 mg/oz =	_____
_____ Tablets Anacin, Vanquish Bromoseltzer, Midol, Darvon, Fiorinal _____	X	30 mg/tablet =	_____
_____ Tablets Excedrin _____	X	60 mg/tablet =	_____
_____ Caffeine "shots" vary from 3 to 200+ mg/fl oz			_____
Total caffeine intake milligrams/day =			_____

Toxic effects (including tachycardia, ventricular arrhythmia, and seizures) are observed at approximately 1,200 mg, or 1.2 g (0.15 tablespoons of caffeine).

A life-threatening dose of caffeine is typically estimated at between
10,000 and 14,000 mg, or 10 and 14 g (1.2 – 1.7 tablespoons of caffeine), although smaller doses can be life-threatening in children or
other sensitive populations).

One teaspoon of a powdered pure caffeine product can contain
3,200 mg (or 3.2 g) of caffeine, or the equivalent of 28 cups of coffee. A half cup of a liquid concentrated caffeine product may contain
2,000 mg of caffeine.

A simple mistake, such as measuring a serving in grams rather than
milligrams, could result in a toxic dose being consumed.

Excerpt from: US Department of Health and Human Services
Food and Drug Administration Center for Food Safety and Applied
Nutrition.

Concentrated Caffeine in Dietary Supplements: Guidance for
Industry. April 2018. College Park, Maryland. https://www.fda.gov/
FoodGuidances

1D Assessment of Fluid Intake

General Recommendation
64 oz (8 cups) total fluid/day is recommended to prevent dehydration
from fluid loss through sweat, lungs, urine, etc.

OR calculated by individual's body weight
    0.5 ounces × Body Weight in Pounds = Daily Fluid Requirement
    in ounces
    Example: 160 lbs × 0.5 = 80 oz = 10 cups/day recommended

OR calculated by individual's calorie intake
    0.034 ounces × Daily Caloric Intake = Daily Fluid Requirement
    in ounces
    Example: 2,100 calories/day × 0.034 = 71.4 oz fluid
    recommended/day

Common measures:

   1 can soft drink = 12 oz
   1 small glass juice = 4–6 oz
   Small coffee = 5–8 oz
   Large coffee = 12–16 oz
   Tall glass of tea = 12–16 oz
   Big Gulps = 30 oz
   1 liter = 33.8 oz
   1 quart = 32 oz

Estimated Fluid Need:_____

Estimated Fluid Intake:_____

1E Assessment of Attitude/Emotions About Eating, Food, Weight

Eating Disorders

Assessment for the presence of an eating disorder (anorexia nervosa, bulimia, etc.) or disordered eating can be conducted by asking questions such as the following:

Do you ever feel extremely guilty after eating?

Do you eat differently alone compared to when you are around others?

Have you ever gone on eating binges where you feel that you are unable to stop?

Are you preoccupied with the thought of having fat on your body?

Have you ever self-induced vomiting to control your weight or shape?

Have you ever used laxatives, diet pills, or diuretics (water pills) to control your weight or shape?

Have you ever exercised more than 60 minutes a day to lose or control your weight?

Disordered Eating

An individual with disordered eating often reports the same behaviors as those with eating disorders but at a lesser frequency or lower level of severity. They may be at risk for developing an eating disorder and are more likely to have a history of, or be at risk for, depression and/or anxiety. Disordered eating may include, but is not limited to, frequent yo-yo dieting, rigid and unhealthy food and exercise regime, or preoccupation with food, body, and exercise that has a negative impact on quality of life.

Orthorexia

Orthorexia is a term used to describe an individual who is pathologically self-righteous and obsessive about eating and food choices. A few questions indicating possible orthorexia include the following:

1. Are you spending more than three hours a day thinking about healthy food?
2. Does your self-esteem get a boost from eating healthy?
3. Do you look down on others who don't eat this way?
4. Does your diet make it difficult for you to eat anywhere but at home, distancing you from friends and family?
5. Do you eat differently when you are alone than when you are eating with others?

Individuals with orthorexia may need to be assessed for anxiety, obsessive-compulsive disorder (OCD), or underlying fears they are trying to control through restriction of food intake and choices.

Ref: Koven NS, Abry AW. The clinical basis of orthorexia nervosa: emerging perspectives. *Neuropsychiatr Dis Treat*. 2015 Feb 18;11:385–94. doi: 10.2147/ NDT.S61665. https://www.ncbi.nlm.nih.gov/ pubmed/25733839

Orthorexia may lead to nutritional deficiencies, medical complications, and poor quality of life. Overlapping symptoms of orthorexia and anorexia nervosa also may include OCD, obsessive-compulsive personality disorder (OCPD), somatic symptom disorder, illness anxiety disorder, and psychotic spectrum disorders.

See also

Academy of Nutrition and Dietetics: http://www.eatright. org/resource/health/diseases-and-conditions/eating-disorders/ what-is-disordered-eating

The Eating Attitudes Test (EAT-26) http://www.eat-26.com/

Yale Food Addiction Scale (YFAS) http://fastlab.psych.lsa.umich. edu/yale-food-addiction-scale/

Body Mass Index – a measure of health considering height and weight

https://www.nhlbi.nih.gov/health/educational/lose_wt/BMI/
bmicalc.htm
https://www.webmd.com/diet/body-bmi-calculator
https://www.cdc.gov/healthyweight/assessing/bmi/adult_bmi

$$BMI = weight/height^2 \quad kilograms/meters^2$$
$$or$$
$$pounds/inches^2 \times 703$$

BMI:

Underweight = < 18.5

Normal weight = 18.5–24.9

Overweight = 25–29.9

Obesity = BMI of 30 or greater

BMI does not take into account a person's gender, age, fluid or electrolyte status, muscular build, or other factors that influence health.

Waist-to-Height Ratio – a measure of weight distribution and health considering waist-to-height ratio: height in inches/waist in inches
https://www.bmi-calculator.net/waist-to-height-ratio-calculator/
#result

Obese = 0.58 and above

Overweight = 0.49–0.57

Healthy = 0.42–0.48

Underweight = Below 0.42

South London and Maudsley Eating Disorder Clinic BMI Guidelines

https://seedeatingdisorders.org.uk/page/bmi

20.0 – 18.5 Normal "Healthy" Weight

18.5 – 17.5 Underweight

17.5 – < 15 Anorexia Nervosa

15.0 – 13.5 Severe Anorexia

13.5 – 12.0 Critical Anorexia: Organs Fail

< 12 Life-Threatening

Additional Aspects of Nutritional Assessment

See also: Table 1 – Reports and Observations Which May be Related to Nutritional Risk

- **Comorbid Health/Mental Health Conditions**
- **Lifestyle Choices and Habits**
- **Genetics**
- **Use of Supplements**
  Form 3. Supplement Intake
  Form 4. Evaluation of Supplement Use
  Table 2. Nutrients and Safe Upper Limits
  Table 3. Megadoses of Nutrients
  Table 4. Potentially Toxic Minerals/Elements Found in Food
- **Nutrient – Medication Interactions**
  Form 5. Nutrient: Medication Interactions
  Table 5. Potential Nutrient: Drug Interactions
- **Laboratory Assessment of Nutritional Status**
  Table 6. Laboratory Assessment of Nutritional Status
  Laboratory Resources – Direct to Consumer Laboratory Testing

Comorbid Health Conditions Occurring Frequently with Mental Health Conditions

- Diabetes
- Hyperglycemia
- Hypoglycemia
- Cardiovascular disease
- Smoking
- Substance abuse
- Socioeconomic factors such as food insecurity

Comorbidity is defined as the co-occurrence of mental and physical disorders in the same person, regardless of the chronological order in which they occurred or the causal pathway linking them.

**Goodell S, BG Druss and ER Walker.
"Mental Disorders and Medical Comorbidity".
The Synthesis Project. The Robert Wood
Johnson Foundation. 2011. Princeton, NJ.**

- Diseases that result in depression and that affect appetite or ability to prepare food (for example HIV, cancer)
- Oral/dental health, pain, inability to chew

- **Lifestyle Choices and Habits**
Some life situations and conditions strongly influence the likelihood of an individual having a nutritionally inadequate diet. Presence of these factors serve as an alert to look more closely when using other assessment tools

   Living situations with limited facilities for food preparation/ preservation
   Frailty/disability – difficulty standing, shopping, food preparation
   Isolation, loneliness – social eating is motivating
   Lack of appetite, nausea, pain
   Decreased sensitivity to taste and flavors

- **Genetics**
A variety of conditions include the risk of a nutritional issue. If the genes linked to these health conditions are inherited, the nutrition risk may also present.
- **Selected Inheritable Conditions with Potential Nutrition Concerns**

   Alcoholism
   Anemia (Pernicious)
   Anorexia Nervosa
   Bipolar Disorder
   Cancer
   Depression
   Diabetes
   Food Allergy
   Galactosemia
   Intestinal Disorders
   Heart Disease
   Hemochromatosis
   Kidney Disease
   Maple Sugar Urine Disease (MSUD)
   Migraine Headaches
   Osteoporosis

Phenylketonuria
Schizophrenia
Thyroid Disorder
Wilson's Disease

Resources for additional information on genetics:

1) Genetics Home Reference – Health Conditions https://ghr.
nlm.nih.gov/condition
2) https://www.genome.gov/For-Patients-and-Families/
Genetic-Disorders
3) https://medlineplus.gov/genetics/condition/
4) https://infocus.nlm.nih.gov/2018/04/25/genetics-home-
reference-turns-15/

Selected examples of additional genetic changes related to nutritional health:

Biotin-thiamine-responsive basal ganglia disease (BTBGD) is recurrent subacute encephalopathy manifesting as confusion, seizures, ataxia, dystonia, supranuclear facial palsy, external ophthalmoplegia, and/or dysphagia, which if left untreated can eventually lead to coma and even death. The diagnosis is related to variants of biallelic *SLC19A3*. Treatment is biotin (5–10 mg/kg/day) and thiamine (in doses ranging from 300–900 mg), given orally as early in the disease course as possible and is continued lifelong. Symptoms typically resolve within days.

Iron-refractory iron deficiency anemia results from a deficiency of iron in the blood. It is described as "iron-refractory" because the condition is totally resistant to treatment with iron given orally and partially resistant to iron given in other ways, such as intravenously (IV).

Hereditary hemochromatosis causes the body to absorb too much iron from the diet, which is stored in the skin, heart, liver, pancreas, and joints. Humans cannot increase the excretion of iron. Early symptoms of hereditary hemochromatosis may include fatigue, joint pain, abdominal pain, and loss of sex drive. Later signs and symptoms can include arthritis, liver disease, diabetes, heart abnormalities, and skin discoloration. The appearance and progression of symptoms can be affected by factors such as the amount of iron in the diet, supplements, alcohol use, and infections.

Hypermethioninemia is an excess of the amino acid methionine in the blood. This can occur when methionine is not metabolized properly. People with hypermethioninemia often do not show any symptoms, but some have intellectual disabilities, neurological problems, delays in motor skills, liver problems, unusual facial features. The breath, sweat, or urine may have an odor resembling boiled cabbage. Hypermethioninemia can occur with other metabolic disorders, such as homocystinuria, tyrosinemia, or galactosemia. It can also result from liver disease or excessive dietary intake of methionine from consuming large amounts of protein or a methionine-enriched infant formula.

The *MTHFR* gene provides instructions for making an enzyme called methylenetetrahydrofolate reductase, which plays a role in processing amino acids, the building blocks of proteins. Methylenetetrahydrofolate converts 5,10-methylenetetrahydrofolate to 5-methyltetrahydrofolate. This reaction is required for the multistep process that converts the amino acid homocysteine to methionine. The body uses methionine to make proteins and other important compounds.

At least 40 mutations in the *MTHFR* gene have been identified in people with homocystinuria, a disorder in which the body is unable to process homocysteine and methionine properly. Several polymorphisms in the *MTHFR* gene have been associated with an increased risk of neural tube defects, a group of birth defects that occur during the development of the brain and spinal cord. Most people with *MTHFR* gene polymorphisms do not have neural tube defects, and their children are also typically unaffected. Changes in the *MTHFR* gene are only one of many genetic and environmental factors that are thought to contribute to these conditions.

See also https://www.ncbi.nlm.nih.gov/gene?cmd=search&term=Tetrahydrafolatereductase

The *PRPS1* gene provides instructions for making an enzyme called phosphoribosyl pyrophosphate synthetase 1, or produce phosphoribosyl pyrophosphate (PRPP) synthetase. This enzyme helps PRPP. PRPP is involved in producing purine and pyrimidine nucleotides, the building blocks of

DNA, RNA, as well as ATP and GTP, energy sources in the cell. PRPP synthetase 1 and PRPP also play a key role in recycling purines from the breakdown of DNA and RNA, a faster and more efficient way of making purines available. Phosphoribosylpyrophosphate synthetase superactivity (PRS superactivity) results in overproduction and accumulation of uric acid in the blood and urine. The overproduction of uric acid can lead to gout, caused by an accumulation of uric acid crystals in the joints, often treated by diet.

Sickle cell anemia is a disease in which red blood cells, that are normally shaped like a disc, become a sickle or crescent shape and become fragile. These abnormal cells deliver less oxygen to the body's tissues. They may lodge in small blood vessels and break into pieces, interrupting blood flow and reduce even more the amount of oxygen flowing to body tissues. People with this condition should take folic acid supplements because they are needed to produce new red blood cells.

Thiamine-responsive megaloblastic anemia syndrome is a rare condition characterized by hearing loss, diabetes, and mega-loblastic anemia. The anemia can be treated with high doses of vitamin B1 (thiamine). Individuals may develop degeneration of the nerves carrying information from the eyes to the brain.

Transcobalamin deficiency impairs the transport of cobalamin (vitamin B12). People with transcobalamin deficiency often develop megaloblastic anemia, a shortage of red blood cells. Individuals may also have a shortage of white blood cells (neutropenia), which can lead to reduced immune system function. Decreased cellular cobalamin can lead to methyl-malonic aciduria or homocystinuria.

Trimethylaminuria is a disorder in which the body is unable to break down trimethylamine, a chemical compound that has a pungent odor, often described as smelling like rotting fish, rotting eggs, garbage, or urine. As this compound builds up in the body, it causes affected people to give off a strong odor in their sweat, urine, and breath, which may vary in intensity over time and can be quite distressing to the victim. For suggestions for management of trimethylaminuria, see https://www.genome.gov/11508983/#al-6.

Warfarin sensitivity is a condition in which individuals have a low tolerance for the anticoagulant drug warfarin. One copy of the altered gene in each cell is sufficient to result in warfarin sensitivity. Different polymorphisms affect the activity of warfarin to varying degrees. Warfarin sensitive people may be at risk of an overdose, which can cause abnormal bleeding in the brain, gastrointestinal tract, or other tissues, leading to serious health problems or death. Patients are advised to maintain a *stable and consistent* intake of vitamin K in their diet and *not avoid* vitamin K-containing foods.

Research on the effect of genes on nutritional issues such as personal taste preferences, metabolism, satiety, and weight change is in its infancy. Some interesting reports include the following:

- Prof. Sadaf Farooqi and colleagues at Cambridge indicate that individuals with the MC4R (Melanocortin 4 receptor) gene defects might unknowingly prefer high-fat foods because of the MC4R brain pathway not functioning, which contributes to them gaining weight.

  Farooqi, I. S. Genetic, Molecular and Physiological Insights into Human Obesity. *Eur J Clin Invest.* April 2011, 41(4): 451–455. doi: 10.1111/j.1365-2362.2010.02468.x. PMID: 21391993

- S. J. Melhorn et al. at the University of Washington in Seattle have shown that the first intron of the fat mass and obesity-associated (FTO) gene increase obesity risk. People with "high-risk" FTO genotypes exhibit preference for high-fat foods, reduced satiety responsiveness, and greater food intake consistent with impaired satiety.

  Melhorn, S. J. et al. *The American Journal of Clinical Nutrition*, Volume 107, Issue 2, 1 February 2018, pp. 145–154, https://doi.org/10.1093/ajcn/nqx029

- Dr. Nicola Pirastu and Dr Antonietta Robino, from the University of Trieste and the IRCCS Burlo Garofolo Institute for Maternal and Child Health, Trieste, Italy, uncovered 17 independent genes related to liking for certain foods, including artichokes, bacon, coffee, chicory, dark chocolate, blue cheese, ice cream, liver, oil or butter on bread, orange juice, plain yogurt, white wine and mushrooms. Surprisingly, none of the genes thus identified belonged to the category of taste or smell receptors. "There is still much that needs to be done... . For example, we found a strong correlation between the HLA-DOA gene and white wine liking, but we have no idea which of the characteristics of white wine this gene influences."

Pirastu, N., A Robino, et al. Genetics of Food Preferences: A First View from Silk Road Populations. *J of Food Science*. August 2012, 77(12): S413.

- **Use of Supplements**
  Some nutrients, though they are essential to the human body, at high levels may be harmful or toxic. The range between necessary and toxic varies. Nutrients in food are rarely at a toxic level. Supplements may contain up to thousands of times the amount recommended for health for the general public. Individual products and the total sum of individual nutrients from supplement products, combined with estimated food sources, should be included in assessment of supplement use.

  Example: A mature female taking several supplements as well as meal/calorie supplement drinks was inadvertently consuming toxic levels of selenium. Each source provided 100% of the daily recommendation intake (DRI) for selenium. One sign that was distressing to her was hair loss, which could also have been attributed to, or caused by, a treatment/medication side effect.

  See also Forms 3 and 4, and Tables 3, 4, and 5.

Tolerable Upper Intake Unit (UL – Upper Limit): The highest level of nutrient intake that is likely to pose no risk of adverse health effects for almost all individuals (98%) in the general population. As intake increases above the UL, the risk of adverse effects increases.

Form 3. Supplement Intake

Name _____ Date_____

It is important to include nutrition in improving your mental and physical health.

Please share the following:

1. Do you take any nutritional supplements? No ___ Yes ___
                    (please circle which supplements you take)

Vitamins	Vit A	B-1(thiamin)	B-2 (riboflavin)
	B-3 (niacin)	B-6 (pyridoxine)	B-12 (cobalamin)
	C (ascorbic acid)	D (ergosterol),	E (tocopherol)
	K (menaquinone)	Folic acid (B-9)	
	Multi-vitamin supplement_____		
Minerals	Calcium (Ca)	Iron (Fe)	Magnesium (Mg)
	Zinc (Zn)	Selenium (Se)	Chromium (Cr)
	Multimineral supplement _____		
	Do you use iodized salt? Yes No Don't know		
Other	Omega-3 Fatty acids	Fish oil	Amino acids _____
	Protein powder	Whey	Casein _____
	Liquid drinks Ensure	Boost	Other _____

Brand name _____

_____

Form 4. Evaluation of Supplement Use

Name_____ Date_____

Include: Vitamins   Minerals   Fatty Acids (O-3)
         Protein Powders Amino Acids   Sports Drinks
         Meal Replacement Drinks (Ensure, Boost, Slim-Fast, etc.)
         Herbs   Spices
         Other _____

SUPPLEMENT – NAME/BRAND	AMOUNT USED	FREQUENCY	EXCEEDS SAFE OR TOXIC DOSE

Provider signature _____ **date**_____

See also
Table 2. Nutrients and Safe Upper Limits
Table 3. Mega doses of Nutrients

Table 2 Nutrients and Safe Upper Limits

NUTRIENT	UL/RISK (ADULTS)	POTENTIAL EFFECT OF EXCESS
Vitamins		
Vitamin A	10,000 IU/d	Liver damage, hair loss, dry skin, headache, joint pain
Vitamin B-6	100 mg/d	Numbness, pain, neurological problems
Vitamin C	2,000 mg/d	Nausea, diarrhea, kidney stones
Vitamin D	4,000 IU/d	Calcification of brain and arteries, appetite down, nausea; may induce deficiency of vitamin K
Vitamin E – tocopherol	1,100 IU –synthetic; 1,500 IU – natural	Decrease in blood clotting time
Vitamin K	1,000 mcg (UK "guidance level")	Anemia
Folic acid (vitamin B-9)	Adults: 1,000 mcg/d DFE (dietary folate equivalent) less for children and adolescents 1 DFE = 1 mcg naturally occurring folate = 0.6 mcg folic acid taken with food = 0.5 mcg folic acid taken on an empty stomach. https://ods.od.nih.gov/factsheets/Folate-HealthProfessional/#h2 http://www.consumerlab.com/RDAs/	Masks B-12 deficiency Can interact with other medications
Niacin	35 mg/d	300 mg/day (recommended for alcohol withdrawal) not likely to cause side effects; 1–2 g/day for hyperlipidemia, can cause flushing side effect.
Minerals		
Boron	20 mg/d	Reproductive system
Calcium	Up to 50 yr of age: 2,500 mg/d 51+ yr of age 2,000 mg/d	Constipation, nausea, kidney stones
Choline	3.5 g/d	Body odor – see trimethylaminuria
Copper	10,000 mcg/d = 10 mg/day	Liver damage, gastrointestinal disturbance Copper overload: increased norepinephrine and lower dopamine (Walsh)
Fluoride	10 mg/d	Bone, nerve, kidney, muscle damage
Iodine	1100 mcg/d	Thyroid hormone up

(Continued)

Table 2   (Continued) Nutrients and Safe Upper Limits

NUTRIENT	UL/RISK (ADULTS)	POTENTIAL EFFECT OF EXCESS
Iron	45 mg/d	Gastrointestinal distress
Lithium	varies with age, weight, symptoms	diarrhea, vomiting, drowsiness, muscle weakness, tremors, unsteadiness, muscle control or coordination (see also mayoclinic.org)
Magnesium	400 mg/d	Diarrhea
Manganese	11 mg/d	Neurotoxicity
Molybdenum	2,000 mcg/d	Joint pains, uric acid up
Nickel	260 mcg/d	Skin sensitivity/reaction
Phosphorous	19–70 yr of age – 4 g/d 70+ yr of age – 3 g/d	Parathyroid hormone altered; bone density down
Selenium	400 mcg/d	Hair and nail loss, nausea, vomiting
Vanadium	Not established	Fatigue, gastrointestinal irritation
Zinc	40 mg/d	Impaired immune function; low HDL
Copper: Zinc ratio	Assaultive subjects Cu: Zn ratio: 1.40 Controls " " " 1.02	Aggressive behavior
Other		
Fish oil		May provide excessive vitamin A or mercury (Hg)
Omega-3 fatty acids	1000 mg/d 06:03 ratio 2:1 to 4:1 EPA:DHA ratio	Caution re excess O-3 supplements used simultaneously with blood thinners such as warfarin or aspirin
Leucine – amino acid	14 mg/kg/day (105 lb person = 68 kg x 14 = 954 mg/day)	May interfere with niacin metabolism, inducing pellagra; may increase insulin resistance; may reduce dopamine levels

Table 3 Megadoses of Nutrients

NUTRIENT	DRI*	UL**	MEGADOSE***
Pyridoxine	1.3 mg	100 mg	1,000 mg
Thiamin	1.1 mg	-	1,000 mg
Riboflavin	1.1 mg	-	400 mg
Niacin	14 mg	35 mg	2,000 mg
Biotin	30 μg	-	100,000 μg
Cobalamin (B$_{12}$)	2.5 μg	-	1,000 μg
Folic acid	400 μg	1,000 μg	40,000 μg
Vitamin K	90 μg	-	45,000 μg
Calciferol (Vit D)	5 μg	50 μg	5000 μg
Tocopherol	15 mg	1,000 mg	800 mg
Tetrahydrobioterin	-	-	40 mg
S-adenosylmethionine	-	-	800 mg
Pantothenic acid	5 mg	-	150 mg
Lipoic acid	-	-	300 mg
Carnitine	-	-	2,000 mg
Thyroid hormone	-	-	1.75 mg
Serine	-	-	500 mg/kg/day
Glycine	-	-	200 mg/kg/day
Zinc	8 mg	40 mg	-
Potassium	2,000 mg	-	-

Table 4 Potentially Toxic Minerals/Elements Found in Food

POTENTIALLY TOXIC: MAY BE FOUND IN FOODS, HERBS, REMEDIES	SOURCES	UNSAFE AMOUNT
Mercury	Fish highest in mercury: Mackerel (king) marlin, orange roughy shark, swordfish, tilefish,tuna (bigeye, ahi)	Daily reference dose (RfD), RfD = ≤ 0.1 μg/kg bodyweight/ per day Recommend: Whole blood mercury level < 5.0 μg/L or hair level < 1.0 μg/g
Selenium	Brazil nuts (80–90 mcg/nut), tuna (90 mcg in 3 oz, other fish half as much or lower)	> 400 mcg/day
Lead	Food-relate sources: lead-glazed pottery, crystal, pewter, imported food cans with lead solder; lead solder in drinking water pipes	Any amount

(*Continued*)

Table 4   (Continued) Potentially Toxic Minerals/Elements Found in Food

POTENTIALLY TOXIC: MAY BE FOUND IN FOODS, HERBS, REMEDIES	SOURCES	UNSAFE AMOUNT
Arsenic	Rice (esp., parboiled, imported), beer (esp. if rice used as ingredient), apple juice	Any amount 10 ppb Children not > 4–5 oz apple juice/day
Glycyrrhizic acid	Black licorice; FDA warns that eating as little as two ounces of black licorice daily for two weeks could lead to heart rhythm problems, especially among people older than 40. Glycyrrhizin, a compound derived from the licorice root, can lead to dropping potassium levels, which can cause abnormal heart rhythms, high blood pressure, edema, https://www.fda.gov/consumers/consumer-updates/black-licorice-trick-or-treat	> 2 oz /day for 2 weeks
Cassia cinnamon/ coumarin (coumarin not present in Ceylon cinnamon)	Cinnamon in human studies is shown to lower blood glucose and lipid levels. *Diabetes Care* 2003 (Dec) 26:12: 3215–3218. https://care.diabetesjournals.org/content/diacare/26/12/3215.full.pdf Coumarin is a chemical compound found in several plants, including cinnamon, that can cause liver damage in large doses. Coumarin is not present in Ceylon cinnamon. The Federal Institute for Risk Assessment concluded that you'd need to take about a gram per day (1,000 milligrams) – for six months or longer to be at risk	1–6 grams cinnamon/day shown to be safe in human research studies The European Food Safety Authority's guideline for coumarin is 0.1 milligrams per day per kilogram of body weight

Form 5. Nutrient : Medication Interactions

Name_____ date _____

Do you take any of the following medications?

NAME (PRESCRIPTION)	DOSE	REGULARLY HOW OFTEN?	OCCASIONALLY HOW OFTEN?
Thyroid			
Omeprezole			
Isoniazid			
Lithium			
Methotrexate			
Zantac			
Statins			
Metformin			
MonoAmine Oxidase Inhibitor			
Antiepileptics: Dilantin, Valproate, Carbamazepine			

NAME (NONPRESCRIPTION; OVER THE COUNTER)	DOSE	REGULARLY HOW OFTEN?	OCCASIONALLY HOW OFTEN?

I don't have/have digestive problems (upset stomach, nausea, vomiting).
How often?_____
I don't have/have constipation, diarrhea.
How often?_____

Assessment of Nutrient – Medication Interactions

The interactions of medications and nutrients may interfere with the absorption, metabolism, or function of either. This may influence physical or mental health in a variety of ways, including effect on appetite and dental health.

Table 5   Potential Nutrient: Drug Interactions

MEDICATIONS WITH POTENTIAL FOR NUTRIENT/DRUG OR DRUG/ NUTRIENT INTERACTION	NUTRIENT INVOLVED
Erythromycin drugs alter intestinal flora and may alter absorption of thiamin	Thiamin
Protein pump inhibitors (Omeprazole)	B-12
Histamine H-2 receptor blockers (ranitidine), metformin, colchicine	B-12
Contraceptives, methotrexate, phenytoin, Depakote, cholestyramine	Folate/folic acid
Barbiturates	Excess selenium may decrease the clearance/excretion time of barbiturates
Antacids	May lower absorption of folate, iron, phosphorous; may raise aluminum, magnesium; take separately from citrus fruit, juice, or calcium citrate
Antianxiety	Limit caffeine to < 400 mg/day; avoid stimulant or sedative herbs; avoid grapefruit juice
Antibiotic	PenV-K may raise potassium, sodium levels, give false positive glucose test, may produce black hairy tongue, oral candidiasis
Antidepressant – tricyclic	Incompatible with carbonated beverages, grape juice; limit caffeine, may increase need for riboflavin; black tongue possible; appetite up for sweets
Antidepressant – monoamine oxidase inhibitors	Limit licorice, avoid tryptophan supplements, limit foods with tyramine
Antipsychotic	Appetite changes, elevated cholesterol, glucose up or down, weight increase; take magnesium supplement separately, may increase riboflavin need
Antiseizure/anticonvulsants	Folate supplement frequently prescribed; folic acid is an antagonist of phenytoin (Dilantin), phenobarbital, methotrexate, and other medications. Appetite changes: May need supplement of calcium, vitamin D, B-1, carnitine

(Continued)

Table 5   (Continued) Potential Nutrient: Drug Interactions

MEDICATIONS WITH POTENTIAL FOR NUTRIENT/DRUG OR DRUG/ NUTRIENT INTERACTION	NUTRIENT INVOLVED
Diet pills	If product mechanism is to decrease fat absorption may interfere with fat-soluble vitamins A, D, E, K; some decrease appetite temporarily and are potentially addictive
Diuretics	Monitor potassium, magnesium levels, avoid natural licorice; caution with supplements of vitamin D and calcium; may interfere with B-6 metabolism
Hypoglycemics (oral)	Metformin may lower B-12, folate, raise homocysteine while lowering lipids
Insulin	May lower potassium, magnesium phosphorous, may raise T-4
Laxative	May lower potassium, calcium; monitor electrolytes with excess use
Lipid-lowering	With statins avoid grapefruit* and its juice; avoid hi-dose niacin, red rice yeast
Lithium	Requires consistent fluid and sodium intake; cases of Li-induced goiter reported in the literature
Methotrexate	Lowers absorption of folate; gingivitis
Selective serotonin reuptake inhibitor (SSRI)	Avoid tryptophan supplement; caution with grapefruit juice
Tobacco	Increases need for vitamin C from 60 mg/day to 100 mg/day
Thyroid	Absorption lowered with iron, calcium, magnesium supplement, with soy milk, soy foods, walnuts

* Grapefruit juice contains furanocoumarins, which block the action of the enzyme CYP3A4, which metabolizes some drugs (example: statins, Lipitor, Procardia). Drinking grapefruit juice can result in buildup and higher blood levels of the drug.

- **Laboratory Assessment of Nutritional Status**

Evaluation of laboratory results or biomarkers (which may be available through the patient's medical record) can indicate nutritional risk.

Laboratory/Biochemistry Indicators of Nutritional Status

Suggested Laboratory Assessment for Nutrients Often Related to Mental Status

Vitamins and Minerals

Vitamin B-12/methylmalonic acid (MMA)

Copper
Folic acid (vitamin B-9)
Pyridoxine (vitamin B-6)
Homocysteine
Vitamin D
Magnesium
Selenium
Zinc
Copper: Zinc ratio

Fats (Lipids)

O-3 Fatty acids (EPA, DHA)
Total Cholesterol

Carbohydrates

Hemoglobin A1c

If history includes alcohol intake, bariatric surgery, hyperemesis gravidum, anorexia nervosa, bulimia, nutrients of additional concern include:

Thiamin (vitamin B-1)
Niacin (vitamin B-3)

See also Table 6. Laboratory Testing of Nutritional Status

Table 6   Laboratory Assessment of Nutritional Status

Biochemical/laboratory assessment, with follow-up to verify change related to treatment, is the most convincing way to diagnose and treat a nutritional deficiency

NUTRIENT	RECOMMENDED LABORATORY ASSESSMENT	REFERENCE RANGE/EXPECTED/N VALUES (SEE ALSO N ESTABLISHED BY INDIVIDUAL LABORATORY USED)	NOTES
Vitamins			
Vitamin B-12 Forms: Cyanocobalamin methylcobalamin Adenosylcobalamin Hydroxocobalamin	MMA: methylmalonic acid	Serum: 0–0.4 µmol/L, 0–4.7 µg/dL Urine: reference interval: 0–3.6 mmol/mol creatinine	
Folic acid: Vitamin B-9 Forms of folate: Folate – in natural foods Folic acid – synthetic, in supplements Folacin – generic term for above terms Folinic acid – a derivative, full activity L-5- tetramethylhydrofolate or 5 formyl tetrahydrofolate – the natural form found in supplements	Folate, folacin, folic acid, FIGLU, tetrahydrofolate reductase (THFR)	Serum 5–25 ng/ml; 11–17 nmol/L RBC 360–1,400 nmol/L	
Vitamin D Forms: D-2 ergocalciferol D-3 cholecalciferol 25-hydroxy-vitamin D – calcifediol; 1,25 dihydroxy-vitamin D – calcitriol	Serum 25(OH)D Serum concentration of vitamin 25(OH) D is the best indicator of vitamin D status.	Recommended: ≥ 30 ng/ml). Deficiency: < 120 nmol/l 48 ng/ml Upper N range 225 nmol/l; 90 ng/ml Toxicity: rare if below 500 nmol/l; 200 ng/ml	Calcitriol is a man-made active form of vitamin D. Calcitriol is used in patients with kidney disease who can't make enough of the active form of vitamin D.

Nutrient	Test	Reference values	Testing laboratories / series
		Vitamin D Council: Deficient: 0–30 ng/ml Insufficient: 31–39 ng/ml Sufficient: 31–39 ng/ml Toxic: >150 ng/ml	Testing laboratories Deficient: 0–31 ng/ml Sufficient: 32–100 ng/ml
Thiamin – Vitamin B-1	Change in activity level of dependent enzyme after a known dose of B-2	Increase in activity of TKA >15% = B-1 deficiency	Testing series: 1. Transketolase (TKA) level 2. Followed by challenge of thiamin pyrophosphate (TPP effect) 3. Repeat TKA: the greater the increase in activity of the transketolase enzyme, the greater the deficiency
B-2 – Riboflavin	Erythrocyte glutathione reductase enzyme activity	Erythrocyte glutathione reductase enzyme activity coefficient > 1.4 = great deficiency Adults urinary excretion: 70–199 µg/g creatinine	
B-3 – Niacin, nicotinamide	Urine	Adults: Urine – excretion of N-methyl nicotinamide 1.6–4.29 mg/g creatinine	
B-6 – Pyridoxine	Erythrocyte transaminase enzyme activity	Erythrocyte transaminase index E-AST <1.5 E-AST 1.9–2.2 marginal status E-AST >2.2=deficiency	
Biotin	Biotinidase enzyme –screening newborns		
Vitamin A	Retinol, retinol binding protein (RBP)	Urine: 163 µg/24 hours Serum vit A >20 µg/dL	

(Continued)

Table 6 (Continued) Laboratory Assessment of Nutritional Status
Biochemical/laboratory assessment, with follow-up to verify change related to treatment, is the most convincing way to diagnose and treat a nutritional deficiency

NUTRIENT	RECOMMENDED LABORATORY ASSESSMENT	REFERENCE RANGE/EXPECTED/N VALUES (SEE ALSO N ESTABLISHED BY INDIVIDUAL LABORATORY USED)	NOTES
Vitamin C – Ascorbic acid		Marginal vitamin C status in males: < 23 mmol/L or 0.40 mg/dL; Above 1.6 mg/dL renal clearance rises abruptly (maximum plasma level); Scurvy: plasma level 7–14 mmol/L or 0.13 – 0.24 mg/dL; Scurvy: whole blood level < 20 mmol/L or 0.30 mg/dL; Depletion of 24 days will result in scurvy. N = 0.20–1.9 mg/dL (Wang); Plasma ascorbate < 0.20 mg/dL; Leukocyte ascorbate < 7 mg/L	Personality changes (fatigue, lassitude, and depression) occurred at whole blood levels of 1.21–1.17 mg/100 ml. Physical changes (signs of scurvy) did not appear until levels had gone lower: down to 0.67–0.14 mg/100 ml. Changes in MMPI scores paralleled changes in serum levels
E Tocopherol	Serum	Adults: 0.47–2.03 mg/dL	
K Phylloquinones		11–12.5 sec prothrombin time	
Minerals			**Minerals/Elements/Electrolytes/ Heavy Metals** Blood or Urine levels of many minerals are not good indicators of body tissue stores
Aluminum		0–6 ng/mL	
Calcium	DEXA scan, Ionized CA	Ionized calcium adults: 9–10.5 mg/dl; 2.25–2.75 mmol/L	
Chromium		Hair: 440 ppm; Urine 1–20 nmol/L	

Copper	Plasma – 1.65 +/–8.6 mmol/1 (*Rükgauer) Adult males: 0.91–1.0 µg/ml Females 1.07–1.2 µg/ml On oral contraceptives: 2.16–3.0 µg/ml	
Iodine, T-3, T-4, TSH	T-4 adult: ~5–12 µg/dL; ~60–154 nmol/L TSH: 2-10 µIU/L; 777	
Iron – total	Adult male: 80–180 µg/dL; 14–32 µIU/L Female: 60–160 µg/dL; 11–29 µIU/L	
Hematocrit (Hct)	Male: 42%–52%; 0.42–0.52 volume fraction Female: 37%–47% or 0.37–0.47	
Total Iron Binding Capacity (TIBC)	250–460 µg/dL; 45–82 µmol/L	
Lead	<10 mcg/dL	
Magnesium	Adult: 1.3–2.1 mEq/L 0.65–1.05 mmol/L	(Hypokalemia may be a better indicator of low magnesium than serum magnesium. Shlamovitz, G. Z., http://www.medscape.org/viewarticle/704606)
Manganese	14.3 +/–11.4 nmol/l *	
Selenium	Blood 0.1–0.34 µg/ml Red cells: 0.23–0.36 µg/ml Plasma –0.80 +/–0.36 mumol/1 (*Rükgauer)	

(Continued)

Table 6 (Continued) Laboratory Assessment of Nutritional Status

Biochemical/laboratory assessment, with follow-up to verify change related to treatment, is the most convincing way to diagnose and treat a nutritional deficiency

NUTRIENT	RECOMMENDED LABORATORY ASSESSMENT	REFERENCE RANGE/EXPECTED/N VALUES (SEE ALSO N ESTABLISHED BY INDIVIDUAL LABORATORY USED)	NOTES
Sodium		Adults: 136–145 mEq/L	
Zinc		Plasma: 115 ± 12 µg/dl	May not be reliable indicators of nutritional status for zinc
		Marginal status:	
		10.7–3 µmol/L;	
		0.70–0.85 µg/ml	
		Neutrophils:	
		108 ±11 µg/10	
		Response of alkaline phosphatase to zinc supplementation	
		Plasma 16.6 +/– mumol/1 (*Rükgauer)	
Carbohydrate			
Fasting glucose		< 100 mg/dL; < 6.1 mmol/dl	< 70 mg/dl: hypoglycemia
2-hour, post-prandial glucose		< 140 mg/dL; < 7.8 mmol/dl	> 250 mg/dl observed to altar mental status
Prediabetes		100–125 mg/dl	
Hemoglobin A1c (Hb$_{A1c}$)		4%–5.9%	
Galactose enzymes or metabolites		18.5–28.5 U/g Hb (units per gram of **hemoglobin**)	
Hypoglycemia		< 50 mg/dl	
Impaired fasting glucose between impaired glucose tolerance		100–125 mg/dl	
		140–199 mg/dl	
Lipids/Fats			
Total cholesterol		< 200 mg/dL; < 5.2 mmol/L	
High density lipoproteins (HDL)		40–59 mg/dL	

Low density lipoproteins (LDL)	< 100 mg/dL; < 2.59 mmol/L		
Triglycerides	Adults: Male: 40–160 mg/dL; 0.45–1.81 mmol/L Female: 35–135 mg/dL; 0.40–1.52 mmol/L		
Essential fatty acids (EFA)			
EPA 0-3	0.51 % (±0.43) % total lipids		
DHA 0-3	1.65 % (±0.67), % total lipids		
DHA Red blood cells	~4% of total lipids (1.9%–7.9%)		
DHA plasma	~3.5% total lipids (1.5%–7.5%)		
Arachidonic acid 0-6	8.84 % (±1.66) % total lipids		
AA: DHA ratio	6.03 (±2.23)		
AA: EPA ratio	23.11 (±11.81)		
Proteins			
Albumin	Adults 3.5–5 g/dL; 35–50 g/L	Adults 3.5–5 g/dL; 35–50 g/L	Not a reliable indicator of protein nutritional status
Prealbumin	Thyroxine-binding prealbumin; transthyretin	Prealbumin: >170 mg/L	
Blood urea nitrogen (BUN)	Adults: 10–20 mg/dL; 3.6–7.1 mmol/L	Adults: 10–20 mg/dL 3.6–7.1 mmol/L	
Homocysteine (Hcy)	4–14 µmol/L	4–14 µmol/L	

(Continued)

Table 6 (Continued) Laboratory Assessment of Nutritional Status

Biochemical/laboratory assessment, with follow-up to verify change related to treatment, is the most convincing way to diagnose and treat a nutritional deficiency

NUTRIENT	RECOMMENDED LABORATORY ASSESSMENT	REFERENCE RANGE/EXPECTED/N VALUES (SEE ALSO N ESTABLISHED BY INDIVIDUAL LABORATORY USED)	NOTES
Amino Acids			
Phenylalanine		Normal blood level for phenylalanine is ~0.8 to 1 mg/dl. The maximum normal level has also been defined as 0.125 mM/L Classical PKU as blood phenylalanine may be defined as > 20 mg/dl. Others use criteria of 4–15 mg/dl. 2–10 mg/dl	Phenylalanine in blood required for N development Necessary for enzyme phenylalanine hydroxylase; a single polymorphism on chromosome 4; If untreated, abnormal development from birth; low IQ
Tryptophan			Precursor of neurotransmitter serotonin Serotonin turnover rate in the central nervous system may be modulated by DHA. A small amount of protein (4% of calories) in a high-carbohydrate meal is sufficient to block any meal-induced increases in the ratio of plasma tryptophan to other amino acids Excess serotonin may result in serotonin syndrome and sign/symptom of confusion, agitation, anxiety, hypomania, or, in severe cases, coma. Autonomic symptoms may include profuse sweating and hyperthermia and/or muscle rigidity

Tyrosine			Becomes essential if phenylalanine is not adequate. Tyrosine can spare 78% of the dietary phenylalanine need. Optimal proportions of dietary phenylalanine and tyrosine have been shown to be 60 phenylalanine:40 tyrosine
Homocysteine	Plasma Urine	Plasma Age 0–30 years: 4.6–8.1 μmol/L Age 30–59 years: 6.3–11.2 μmol/L (males); 4.5–7.9 μmol/L (females) Age > 59 years: 5.8–11.9 μmol/L Urine homocysteine: (varies with the technique used) Reference range (24-hour urine collection): 0–9 μmol/g creatinine.	

Other

Analysis of the gut microbiome See also Appendix for selected references	Fecal sample		Psychobiotics

* M Rükgauer, J. Klein, J. D. Kruse-Jarres. Reference Values for the Trace Elements Copper, Manganese, Selenium, and Zinc in the Serum/Plasma of Children, Adolescents, and Adults. *J Trace Elem Med Biol*. 1997 Jun;11(2):92–8. doi: 10.1016/S0946-672X(97)80032-6. PMID: 9285889

Laboratory Resources – Direct to Consumer Laboratory Testing

Some patients may be interested in laboratory testing for nutritional status beyond what has been completed during other health care, including blood or cell levels of nutrients, gut/fecal microbiome analysis, and genetic analysis. Direct to Consumer Labs provide an assortment of such services. An online search for "Direct to Consumer Laboratory assessment" will provide food for thought on this topic.

The following are examples of laboratories that may perform tests requested by an individual or may be ordered through a health care professional. State laws vary.

Direct Labs is a direct-to-patient testing service available through Quest Diagnostics. Laboratory tests are ordered online after establishing an account. Locations for blood draws are found on the website. **http://www.directlabs.com/**

Spectra Cell Laboratories, Inc., is a CLIA accredited clinical laboratory that specializes in micronutrient testing (MNT), based on 18 years of research at the University of Texas. The micronutrient tests measure how micronutrients are actually functioning within white blood cells. It also offers a specialized profile of homocysteine, lipids, and proteins to assess cardiovascular risk. Results with interpretation are provided. The web site provides a resource to locate their draw sites. https://www.spectracell.com/clinicians/products/mnt/ or Spectra Cell Laboratories https://www.spectracell.com/micronutrient-test-panel

Microbiome Labs https://microbiomelabs.com/home/education/

Quest Diagnostics https://www.questdiagnostics.com/home/

Oxford Diagnostic Laboratories is a division of Oxford Immunotec based in Memphis Tennessee. It performs the T-SPOT® tests for tuberculosis. http://www.oxforddiagnosticlabs.com/home/

Vitamin D Council – In addition to information about vitamin D, the Vitamin D Council offers an in-home test kit that can be ordered by consumers for assessing 25(OH)D^1 vitamin D status. It involves putting a spot of blood on the blood spot card mailed to you and mailing it back. A second option is for individuals to order a test online and get a blood draw at a laboratory close by.

Note

1 25(OH)D is said to be preferred over a test for 1,25(OH)$_2$D https://www.vitamindcouncil.org/

Bibliography

Elmadfa, I and Meyse, A. L. Developing suitable methods of nutritional status assessment: a continuous challenge. *Adv Nutr.* 2014 Sep;5(5):590S–598S.   https://www.ncbi.nlm.nih.gov/pubmed/?term= ibrahim+elmadfa+developing+suitable+methods+of+nutritional+status+-assessments

Jones, A.D. Food Insecurity and Mental Health Status: A Global Analysis of 149 Countries. *Am J Prev Med* 52; 2017. DOI: http://dx.doi.org/10.1016/j.amepre.2017.04.008

Koulouri, Olympia Gurnell, Mark Gurnell, How to Interpret Thyroid Tests. *Clinical Medicine* 2013, Vol 13, No 3: 282–286.

National Research Council (US) Subcommittee. Chapter 6. Protein and Amino Acid. *Recommended Dietary Allowances* Tenth Edition. Washington (DC): National Academies Press (US); 1989. http://www.advancedmolecularlabs.com/pre-workout-studies/the-optimal-timing-of-leucine-consumption/

Tabarki, Al-Hashem, and Alfadhel. Biotin-Thiamine-Responsive Basal Ganglia Disease. *Gene Reviews.* 2013 https://www.ncbi.nlm.nih.gov/books/NBK169615/11/21/2013

Walsh, W.J., Glab, L.B., and Haakenson, M.L. Reduced violent behavior following biochemical therapy. *Physiol Behav.* 2004 Oct 15;82(5):835–839.

Walsh, William J. *Nutrient Power: Heal Your Biochemistry and Heal Your Brain.* 2012. Skyhorse Publishing. New York.

Walsh, William J., et al. Elevated Blood Copper/ Zinc Ratios in Assaultive Young Males. *Physiology & Behavior.* 1997, 62 (2):327–329.

Zaslavsky, Oleg, Zelber-Sagi, Shira, Hebert, James R., et al. Biomarker-calibrated nutrient intake and healthy diet index associations with mortality risks among older and frail women from the Women's Health Initiative. *Am J Clin Nutr* 2017. http://ajcn.nutrition.org/content/early/2017/04/18/ajcn.116.151530.abstract

III

REFERRALS – PROFESSIONAL COLLABORATION

Referral to a Registered Dietitian Nutritionist

Follow-up or in-depth nutritional assessment or care calls for a referral to a registered dietitian nutritionist (RDN) with experience in behavioral health nutrition or a certified nutrition specialist (CNS).

RDNs must have earned a master's degree in nutrition and associated sciences, completed a one-year approved internship, and passed a national examination to qualify for the credential RDN. Seventy hours of approved continuing education is required to maintain the credential for five-year intervals.

Specialty areas of the Behavioral Health Nutrition (BHN) Dietetic Practice Group of the Academy of Nutrition and Dietetics include (1) mental health, (2) eating disorders, (3) substance abuse disorders, and (4) developmental and/or genetic disorders. RDNs may be available through private practice, as consultants for outpatient treatment clinics or day programs, or as staff members for inpatient treatment facilities. The national or state dietetic association website may be helpful in locating an RDN near your patient: https://www.eatright.org/find-an-expert.

With training and accreditation similar to that of RDNs, CNSs are individuals who have a master's or higher degree with coursework in biochemistry, physiology, anatomy, and clinical or life science with at least nine graduate credits in nutrition. CNSs complete at least 1,000 hours of supervised practice and pass a written examination. Those who earn the CNS credential must complete 75 continuing education credits every 5 years. They help clients or the public reach

DOI: 10.1201/9781003539155-3

health-related goals by suggesting the right eating habits to improve nutrition and overall health using the latest nutritional research.

A "nutritionist" title is a general term, without specified qualifying education or experience.

When an RDN or CNS observes or suspects a nutrient-related medical condition or frank malnutrition (see the following section), referral to the individual's primary care physician for follow-up is made, along with recommendations for nutritional follow-up. When a patient exhibits behavior that suggests a mental health problem, an RDN or CNS refers the patient to a licensed mental health care provider for assessment and care.

A Medical Nutrition Therapy (MNT) Act was introduced in the U.S. Senate in the fall of 2020 (S-4504). The act extends Medicare Part B coverage to diseases and conditions related to unintentional weight loss. One aspect of the MNT Act is that it authorizes nurse practitioners, physician assistants, clinical nurse specialists and psychologists to refer their patients for MNT. (watch for developments related to this)

The Body of Knowledge of Clinical Dietetics

Referral to an RDN can be expected to cover the seven domains of the Metaparadigm of Clinical Dietetic, which organizes the body of knowledge used by clinical dietitians in practice, theory, science, and research.

Reference Person – includes the standards, "normal," criteria, used as the basis for comparison, diagnosis, and individualized treatment.

Human Condition – includes the physical and mental condition(s) of the individual who presents for nutritional care.

Practitioner Action and Attitude – includes the behaviors, knowledge, understanding, emotions, beliefs, of the individual RDN who is treating the client.

Practitioner Environment – includes physical space, social environment, colleagues, public environment.

Client Actions and Attitude – includes behavior, knowledge, understanding, acceptance, emotions, beliefs, etc.

Client Environment – includes family, living situation, lifestyle, habits, physical environment.

Nutraceuticals – includes foods, beverages, supplements of calories, protein, fats, carbohydrates, vitamins, minerals, herbs, spices, and other nutritional products.

Assessment of Nutrition Status

The Nutrition-Focused Physical Examination

A Nutrition Diagnosis and Treatment Plan

Interventions by RDNs include recommendations regarding the following:

I. Behavior Change (Client Actions Domain)
   A. Balance activity level with calorie intake
   B. Addition of Supplements, nutraceuticals, food types, activity
   C. Increase of food typically consumed
   D. Decrease use of specific foods, types, etc.
   E. Omission of specific or types of foods, or supplements

II. Intellectual/Emotional Change (Client Attitude domain)
   A. Acceptance
   B. Understanding
   C. Knowledge
      Exposure to content
      Application of knowledge
   D. Move from one stage of readiness to another stage of readiness (Stages of Change Theory/Conceptual Model)
   E. Worldview regarding health, diet, nutrition, cultural factors

III. Environment (Client Environment Domain)
   A. Support System
      Parents
      Spouse
      Friends/coworkers
      Health-care professional – referrals
         Social worker
         Case manager
         Psychologist/family therapist
         MD/psychiatrist
         Home health care
   B. Living circumstances
      Institution
      Self-care vs. cared for by knowledgeable others
      Transportation/food preparation facilities
   C. Finances/food availability
      Treatment Follow-Up
      Nutrition Notes

The Nutrition-Focused Physical Examination

If training, time, space, and equipment are available, the potential for poor nutritional status can be detected with a nutrition-focused physical examination (NFPE). Many RDNs are trained and equipped to conduct the NFPE. Documentation of observations (often with consented medical photographs) is recommended. Verification with laboratory testing and a record of changes that occur with treatment are also recommended serving as validation of the diagnosis and efficacy of treatment.

A NFPE includes physical examination of the oral cavity, eyes, skin, hair, and nails; muscle wasting or development; and girth, noting any observations that indicate possible nutrition insufficiency. Special attention can then be paid to any factors in the individual's health or genetic history that might be causative. Positive findings are most often followed by further laboratory assessment and/or referral.

See also: Appendix for References for the Nutrition-Focused Physical Examination

Form 6. Nutrition Physical Examination Observation Record

Client ID: _____ **Date** _____

BODY AREA	SIGNS	YES	NO	OBSERVATIONS/ NOTES
Lips Mucous membranes	Bilateral angular lesions Desquamation ulceration			
	Scars: pink/white			
	Cheilosis: red			
	Swollen, puffy			
	Fissuring			
	Pallor – lips			
	Pallor – mucous membranes			
Tongue	Red, scarlet, beefy			
	Magenta, purplish			
	Geographic			
	Furrows, scrotal			
	Erosions			
	Lobulated, atrophy			
	Serrations/swelling			
	Fissures			
	Ulcers			
	Red Tip			
	Adhesive: slight			
	Moderate			
	Strong			
	Red lateral margins			
	Adhesive: slight			
	Moderate			
	Strong			
	Filiform papillary atrophy			
	25% 25%–75% 75%			
	Papillary hypertrophy			
	25% 25%–75% 75%			
	Aphthous-like ulcers			
	Dorsum: white / gray / green			
	Under surface: white / gray / green			

(Continued)

Form 6. Nutrition Physical Examination Observation Record

BODY AREA	SIGNS	YES	NO	OBSERVATIONS/ NOTES
Teeth / Gums	Gingival marginal redness			
	Swelling			
	Swollen red papillae			
	Area: small / diffuse			
	Atrophy of interdental papillae			
	Recession of gums			
	Enamel: white / brown			
	Extensive caries			
	Missing teeth			
Eyes	Xanthelasma			
	Angular palpebritis			
	Palpebral petechiae			
	Conjunctiva: Pale			
	Dull			
	Dry			
	Rough			
	Wrinkled			
	Vascularization			
	Sclera: Yellowish			
	Brownish			
	Excessively blue			
	Bitot's spots			
	Cornea: Dull, milky			
	White / gray / brown ring			
	Scars			
	Horizontal band			
	Vascularized			
	Eyelid: Drooping			
	Swollen			
Face and Neck	Moon face			
	Ear lobe crease			
	Nasolabial seborrhea			
	Pallor			
	Light in center brow			
	Dark over cheeks / eyes			
	Dark around neck			
	Enlarged parotid gland			
Hair	Dull			
	Dry			
	Brittle			

(Continued)

Form 6. Nutrition Physical Examination Observation Record

BODY AREA	SIGNS	YES	NO	OBSERVATIONS/ NOTES
Skin / nails	Wire-like, steely			
	Straight			
	Depigmented			
	Sparse / silk			
	Sparse, scaly			
	Flag sign			
	Pluckability			
	Erythema			
	Hyperpigmented areas			
	Red / swollen / blistered			
	Mosaic dermatosis			
	Xerosis			
	Petechiae / purpura			
	Ecchymoses			
	Follicular hyperkeratosis			
	Perifolliculosis			
	Koilonychia			
	Transverse ridging			
	Paronychia			
	Splinter hemorrhages			
	White spots			
Subcutaneous	Tenting test			
	Pitting edema			
Other	Stocking / glove numbness			
	Reflexes: Achilles tendon			
	Knee			

Additional Observations/Notes

_____

_____

Form 7. Assessment of Nutritional Status

This is a suggested form for the comprehensive assessment of nutritional status. Noting these nutritional and psychological descriptors has potential use in patient care for creating a treatment plan and in research for discovering and quantifying the links between nutritional status and mental status. Some items may be omitted in relation to the population being assessed (for example, many individuals will not need to be assessed for phenylalanine or related metabolites or enzymes).

It has been designed for ease in creating information and data that may be transformed to digital form (for example numerical and Yes/No answers that address a single issue).

Patient name _____ M F Age_____

Assessment of Nutritional Status (ANS) (p. 1 of 3)
(Circle any that apply; fill in any known values)

ANS Aspect 1. Risk factors - Family History

Alcohol	Anemia	Anorexia nervosa	Bipolar disorder
Cancer	Depression	Diabetes	Food allergy
Intestinal disorder	Heart Disease	Hemochromatosis	Kidney disease
Migraine Headaches	Osteoporosis	Thyroid disorder	_____

Gene analysis polymorphism (describe) _____

ANS Aspect 2. Physical Status / Body Composition (Circle and/or Fill in Blanks)

Height: _____   Weight: _____   BMI: _____

Yes   No   BMI below 18.5
Yes   No   BMI above 30
Yes   No   Wt. Gain/Loss of ___lbs .... Loss of ____lbs in
                 past ___months.
                 Loss of 10% of weight in 6 months is clinically significant.
Yes   No   Muscle-Wasting
% body fat _____   Yes No   Below 20% – Females / 10% – Males
                          Yes No   Above 35%

ANS Aspect 3. Dietary Habits

a. _____ Eats fewer than three times a day
b. _____ Makes food choices that do not meet the Food Guide Pyramid recommendations
        Yes No 6–11 servings starches   Yes No 2–3 3-oz servings meat/substitute
        Yes No 3–5 servings vegetables   Yes No 2–3 servings fruit
        Yes No 2–3 servings dairy foods Yes No Eats mono-/polyunsaturated fats
        Yes No Not over 10% calories from sugar
        Yes No Not more than (F) 1 (M) 2 drinks alcohol/day
        Yes No Low to moderate use of salt
c. _____ Consumes more than 400 mg caffeine/day
d. _____ Uses nutrient supplements:
        Yes   No   Less than 100% DRI _____
        Yes   No   About or equal to DRI _____
        Yes   No   More than 500% dietary reference intakes (DRI) or greater than
                     UL _____

ANS Aspect 4. Laboratory / Biochemical / Metabolic (above or below normal (N) range for laboratory/biochemistry tests; enter lab value and N value used for comparison)

Carbohydrate
___ Fasting Blood Glucose (FBS) _____   ___ 2-hour postprandial glucose (2 hr PP) _____
___ Hemoglobin A1c (Hb_{A1c}) _____   ___ Galactose – enzymes and/or metabolites____

Assessment of Nutritional Status (ANS) (p. 2 of 3)

Patient name _____ M F Age_____

Lipids
___ Total Cholesterol _____ ___ High Density Lipoprotein _____
___ Low Density Lipoproteins _____ ___ Triglycerides _____
___ Essential Fatty Acids (EFA) and/or Metabolites (EPA, DHA, O-3, O-6) _____

Proteins and Amino Acids
___ Albumin _____ ___ Pre-albumin _____ ___ BUN _____
___ Homocysteine _____ ___ Phenylalanine–related enzymes and/or metabolites _____
___ Other _____

Vitamins (Blood, Serum levels, or Vitamin-Dependent Enzyme)
___ B$_1$ (Thiamin) (TKA) _____   ___ B$_2$ (Riboflavin) _____
___ B$_3$ (Niacin) (Nicotinamide) _____   ___ B$_6$ (Pyridoxine) _____
___ Biotin _____   ___ B$_{12}$ (Cobalamin) (MMA) _____
___ Folacin (Folic Acid) (FIGLU) ____   ___ A (Retinol) _____
___ C (Ascorbic Acid) _____   ___ D (Cholecalciferol) (Ergosterol) _____
___ E (Tocopherol) _____   ___ K (Phylloquinones) _____

Minerals, Elements, Electrolytes, and Heavy Metals
___ Aluminum _____   ___ Calcium, DEXA scan _____
___ Chromium _____   ___ Copper _____
___ Iodine, T-3, T-4 _____   ___ Iron, Hct, TIBC, Hemoglobin, MCV ____
___ Lead _____   ___ Magnesium _____
___ Mercury _____   ___ Potassium _____
___ Selenium _____   ___ Sodium _____
___ Other _____   ___ Other _____

ANS Aspect 5. Clinical Signs and Symptoms (Presence of nutrient-based lesions determined by physical examination (a–e) and/or other symptoms reported by client (f–g))
a. Oral Tongue Lips Gums Teeth _____
b. Skin _____
c. Nails _____
d. Eyes _____
e. Hair _____
f. Yes No   Diarrhea (more than two loose bowel movements/day)
g. Yes No   Constipation (fewer than one bowel movement every three days)
h. Yes No   Dental pain or discomfort that influences eating

ANS Aspect 6. Nutrient ↔ Drug Interaction (Potential for Nutrient/Drug or Drug/ Nutrient interaction) (Check those used, enter drug name if known)
___ Antacids _____   ___ Antianxiety _____
___ Antibiotic _____   ___ Antidepressant _____
___ Antidepressant (Tricyclic) __   ___ Antidepressant (MonoAmine Oxidase Inhibitor) _____
___ Antipsychotic _____   ___ Anti-seizure _____
___ Diet pills _____   ___ Diuretics _____
___ Hypoglycemic (oral) _____   ___ Insulin _____
___ Laxative _____   ___ Lipid-lowering _____
___ Lithium _____   ___ Methotrexate _____
___ Tobacco _____   ___ Thyroid _____
___ Other _____   ___ Other _____

Assessment of Nutritional Status (ANS) (p. 3 of 3)

Patient name _____   M F   Age_____

Nonspecific Signs or Symptoms Reported by Client: (circle any reported; add any additional symptoms)

Appetite ↓ ↑	Concentration reduced	Energy level reduced / increased
Fatigue	Headaches	Irritability
Memory Problems	Sleep Problems	Tearful
_____	_____	_____

Additional Nutritional Observations, Comments:

Summarize findings of the ANS by listing the factors contributing to the determination of an individual's overall Stage of Nutritional Injury.

ANS 1: 0–0.9 = Risk of nutritional injury _____

ANS 2. _____

ANS 3. _____

ANS 4. _____

ANS 5. _____

ANS 6. _____

Nonspecific signs and symptoms_____

ANS: Stages of Nutritional Injury

The Stage of Nutritional Injury (a descriptor of nutritional status) may be assigned to each individual based on any or all of the findings from the assessment and the professional judgment of the practitioner.

Use the following descriptions to determine the Stage of Nutritional Injury of the individual assessed. The highest-level present is most often the designated Stage of Nutritional Injury.

Stages of Nutritional Injury

I. Depletion of nutrient stores, adaptation (potential indicated by ANS Aspects 1 and 4)

II. Reserves exhausted (potential: Stage I indicators of depletion or excesses lasting for six weeks or longer)

III. Physiologic and metabolic alterations (indicated by ANS Aspect 2)

IV. Nonspecific signs and/or symptoms (potential indicated by reports of fatigue, headaches, loss of appetite, decrease in attention, insomnia, etc.)

V. Illness or specific signs and/or symptoms (Indicated by ANS Aspects 3 and 5)

VI. Damage irreversible or nonresponsive to treatment (Potentially including but not limited to loss of absorption sites resulting from bariatric surgery, bone loss, vision loss, loss of nerve function)

Example:
ANS 1: Family hx of alcohol addiction
ANS 2. Loss of 25# past 2 months;
ANS 3. 2 meals/day; 1000 mg caffeine/day

ANS 4. Triglycerides elevated
ANS 5. -0-
ANS 6. Black hairy tongue

Stage of Nutritional Injury: _____ (0–VI)
Nutrition Diagnosis: _____

Form 8. Nutrition-Focused Physical Examination Observations – Short Form

Client ID: _____ Date _____

Clinical Signs and Symptoms
(Presence of nutrient-based lesions determined by physical examination (a–e) and/or other symptoms reported by client (f–g))

a. Oral: Tongue
      Lips
      Gums
      Teeth

_____

b. Skin

_____

c. Nails

_____

d. Eyes

_____

e. Hair

_____

Other Notes

_____

_____

_____

_____

_____ , _____ _____
    Name of examiner   Credential   Date

Form 9. Permission to Photograph

I, _____, grant permission

to _____

to photograph (who, area)_____

_____

for purposes of planning and evaluating nutritional treatment and for educational purposes.

I understand that I will not be personally identified in any professional educational presentations or published collections of these photographs.

_____

Client Signature   Date

or   _____

Signature of Responsible Person;
Relationship   Date

_____

Signature of Clinician   Date

Follow-up photographs
Date_____
Date_____

Diagnosing Malnutrition

Specific medical codes are used for the diagnosis and treatment of malnutrition. A diagnosis of malnutrition has been defined in the *Journal of Parenteral and Enteral Nutrition*.

Resource References:
Phillips, Wendy MS, RD, CNSC. **Coding for Malnutrition in the Adult : What the Physician Needs to Know**

Tappenden, Kelly A., Beth Quatrara, Melissa L. Parkhurst, Ainsley M. Malone, Gary Fanjiang and Thomas R. Ziegler. **Critical Role of Nutrition in Improving Quality of Care: An Interdisciplinary Call to Action to Address Adult Hospital Malnutrition**. *JPEN J Parenter Enteral Nutr* 2013 37: 482 originally published online 4 June 2013

White, Jane V., Peggi Guenter, Gordon Jensen, Ainsley Malone, Marsha Schofield, Academy Malnutrition Work Group, A.S.P.E.N. Malnutrition Task Force and the A.S.P.E.N. Board of Directors. **Consensus Statement: Academy of Nutrition and Dietetics and American Society for Parenteral and Enteral Nutrition: Characteristics Recommended for the Identification and Documentation of Adult Malnutrition (Undernutrition)**. *JPEN J Parenter Enteral Nutr* 2012 36:275.

Xu, Y.C., Vincent, J.I. **Clinical Measurement Properties of Malnutrition Assessment Tools for Use with Patients in Hospitals: A Systematic Review**. *Nutr J* 19, 106 (2020). https://doi.org/10.1186/s12937-020-00613-0

Malnutrition Screening Tools
*Malnutrition Screening Tool (MST)
The MST is a validated tool to screen patients for risk of malnutrition. The tool is suitable for a residential aged care facility or for adults in the inpatient/outpatient hospital setting. Nutrition screen parameters include weight loss and appetite.
*Malnutrition Universal Screening Tool (MUST)
The MUST is a validated screening tool suitable for adults in acute and community settings. The tool is available in several languages.

*Mini Nutritional Assessment (MNA) – Short Form

The MNA® is a quick and easy-to-use screening tool that takes less than 5 minutes to complete. Calf circumference can be substituted for body mass index (BMI) in patients who cannot be weighed or measured. It is the most commonly used form of the MNA®.

*Nutrition Risk Screening (NRS-2002)

This tool includes an assessment of recent weight loss (%), recent intake, BMI, severity of disease, and age.

*Seniors in the Community: Risk Evaluation for Eating and Nutrition (SCREEN©)

SCREEN© is a nutrition screening tool that caters to seniors living in different settings. The questionnaire has undergone significant validation and reliability testing, and is recommended by several national and international bodies, including the World Health Organization.

Sample Nutrition Referral #1
Sample Nutrition Referral

Referral to _____,
Registered Dietitian Nutritionist

From: _____

Referral: Ms. "X"_____

Referral re: ___Constipation _____

Hx: Pt. in counseling for depression and re-normalizing lifestyle; living situation/routines changed from living with partner to living singly. Reports severe, painful, constipation of several months, interfering with life activities.

Nutrition Assessment and Recommendations:

*Height: 5'4 inches   **Weight**: 130 lbs   **Body Mass Index**: 22.5 WNL Patient reports no issues with weight.

*Nutrition-focused physical examination: No physical signs/symptoms of vitamin/mineral excess or deficiency observed or reported.

*Laboratory tests: recent panel all WNL

*Medications: Takes occasional OTC Miralax; Pt. does not want to continue this

*Supplements: Pt. takes moderate-level Vit/Min supplement on a regular/daily basis. (Centrum)

*Recent routine of irregular timing of meals/snacks; eats at fast-food restaurants five to six times a week; reviewed/discussed with patient

> **Food intake** assessed to be very low in fiber < 10 grams/day)
> NIH Recommendation: 20–25 grams fiber per/day for adult women
> **Fluid intake** estimated 45 oz./day
> National Academy Science recommendation for daily fluid intake for women:
> 11–12 cups (88 oz) fluid/day
> **Caffeine intake** ~ 350 mg/day in diet coke and coffee
> **Reported Alcohol intake**: 1 can of beer/day; Coors Light – 0.5 oz alcohol/12 oz can

Recommendations / Plan

*Discussed desired intake and sources, and her preferred foods and beverages.

*A 3-step, 3-week, gradual change in eating pattern planned with Pt. input and agreement.

*Pt. provided with written cc of plan and list of foods to include in daily diet.

Follow-up: Appt in 2 weeks to review progress with eating pattern, food choices and status of constipation.

Signed _____ date _____

Sample Nutrition Referral #2

Sample Nutrition Referral

Referral to _____, Registered Dietitian, RDN

From: _____

Referral: ____Ms. "Y" _____

Referral re: Eating Disorder : Anorexia Nervosa _____

Hx: 19-yr-old female, gradual weight loss for past 8 months, began college 1 year ago; lives in dormitory, referred by parents for counseling and nutrition education; Pt. amenable to learning more about nutrition.

Nutrition Assessment and Recommendations:

* Ht: 5'6"   Weight: 110# BMI: 17.6
        Healthy BMI: 18.5–24.9; Underweight = BMI <18.5

*Nutrition-Focused Physical Exam: Slight muscle wasting; minimal fat deposition;
        Tongue reddened and sensitive to touch;

*Laboratory tests:

Serum folate : 3.4 ng/ml (N= 5–25 ng/ml /11–57 nmol/L)
RBC folate : 300 nmol/L (N= 360–1,400 nmol/L)
Urinary MMA (Methyl Malonic Acid): 2.9 mml/mmol creatinine
        (N= < 3.6 mml/mmol creatinine.)

*Recent routine reviewed/discussed with patient; Eating habits: 1 meal a day, often fast food; frequent snacks (diet soda, sugar-free cookies, occasional fresh fruits);
        Caffeine intake: ~500 mg/day;
        Alcohol intake: Occasional (1–2 drinks, 5–6 x/year)
        Activity level: 1 hour routine/day at the gym: lifting weights, exercise bike, running the track

* Supplements: Consumes a variety of sports drinks on daily basis;
* Assessment and Recommendations
        Underweight: Goal 122 lbs = 55.3 kg; BMI of 19–20
        Calories needed: 2,500 calories/day
        Protein needed: 0.8 g x 55.3 kg = minimum of 44 grams protein/day.
            0.8 g/kg or 0.36 g/lb of wt. (1.1g/kg to gain muscle; 10%–30% of calories)
        Fluid intake: 6–8 cups (55–65 oz) fluids /day) (continued)

* Discussed – Pt. agreeable to goals for protein intake; expressed reluctance to meet goals for calorie intake.
* Plan – Pt. agreed to keep 1 week of written record of protein intake; Agreed to refrain from weighing herself for the coming week

* Pt. provided with a blank food record form and information re protein content of foods.

Follow-up: 1 week

Signed _____, RDN _____
(date)

Nutrition Referral

Referral to _____, Registered Dietitian, RDN   Date _____

From: _____

Patient: _____

Referral re: _____

Hx: _____

Nutrition Assessment and Recommendations:

Ht: _____   Weight: _____   BMI (Body Mass Index): _____
Healthy BMI: 18.5–24.9;

*Nutrition-Focused Physical Exam:

*Laboratory test results :

*Recent routine
Caffeine intake: _____ Alcohol intake/History_____
Fluid Intake: _____

* Supplement Use:

*Medications:

*Relevant Family History, Genetic factors

Activity / Exercise:

* Assessment & Recommendations: _____
* Plan _____
* Pt provided with _____

Follow-up:

Signed _____, RDN _____(Date)

Links to Images of Nutrition-Related Health Problems

http://www.nejm.org/doi/full/10.1056/NEJMicm1403210
Myxedema – Severe Hypothyroidism

http://www.nejm.org/doi/full/10.1056/NEJMicm1406572
Lindsay's Nails – Chronic Kidney Disease

http://www.nejm.org/doi/full/10.1056/NEJMicm1411222
Strawberry Tongue – B-12, Folate

http://www.nejm.org/doi/full/10.1056/NEJMicm1413715
Oral Manifestation of Crohn's Disease

http://www.nejm.org/doi/full/10.1056/NEJMicm1502932
Geographic Tongue

http://www.nejm.org/doi/full/10.1056/NEJMicm1508730
Acanthosis Nigricans and Insulin Resistance

http://www.nejm.org/doi/full/10.1056/NEJMicm1615499?query=
TOC
Nystagmus from Wernicke's Encephalopathy – with Video

http://www.nejm.org/doi/full/10.1056/NEJMicm1207495
Bulimia Nervosa – Dental Erosion

http://www.nejm.org/doi/full/10.1056/NEJMicm1213006
Bilateral Earlobe Creases – High Cholesterol

http://www.nejm.org/doi/full/10.1056/NEJMicm1205309
Bitot's Spot in Vitamin A Deficiency

http://www.nejm.org/doi/full/10.1056/NEJMicm1205540
Severe Vitamin D Deficiency – Rickets

A Colour Atlas and Text of Diet-Related Disorders, 1992; 2nd edition, Donald S. McLaren, MD, PhD, Mosby Yearbook Europe Limited, London.

A Colour Atlas of Nutritional Disorders, Donald S. McLaren, MD, PhD. 1981 Wolfe Medical Publications

Atlas of Clinical Diagnosis, 1995 M. Afzal Mir, DCH, FRCP. W. B. Saunders Company Ltd.

Links to Articles Regarding Nutrition in the *New England Journal of Medicine*

Anemia of Inflammation	*N Engl J Med* 381: 1148–1157 September 19, 2019
Anorexia Nervosa	*N Engl J Med* 382;14 nejm.org April 2, 2020
Atrophic Glossitis	*N Engl J Med* 381: 1568 October 17, 2019
Biotin Mimicking Graves' Disease	*N Engl J Med* 375: 704–706 August 18, 2016
Bitot Spots	*N Engl J Med* 379; 9 nejm.org August 30, 2018
Bulimia Nervosa	*N Engl J Med* 368: 1238 March 28,
Candida; Esophagitis	*N Engl J Med* 376: 1574 April 20, 2017
Celiac Disease and Psychosis	*N Engl J Med* 374: 1875–1883 May 12, 2016
Earlobe Crease	*N Engl J Med* 368: e 32 2013 June 13, 2013
Eating Disorder Not Otherwise Specified	*N Engl J Med* 367; 2 nejm.org July 12, 2012
ER Visits re Adverse Effects of Supplements	*N Engl J Med* 373; 16 nejm.org October 15, 2015
Excess Vitamin A	*N Engl J Med* 374; 9 nejm.org March 3, 2016
Hi Chol, HDL, Xanthomas – Palms	*N Engl J Med* 378 e 26 May 10, 2018
Hypothyroidism – Brittle Nails, Hair Loss	*N Engl J Med* 379: 1363 October, 2018
Hypothyroidism – Ankle Reflex	*N Engl J Med* 379: e 23 October, 2018
Intermittent Fasting	*N Engl J Med* 382: 2541–2551 December 26, 2019

Koilonychia	*N Engl J Med* 379; 9 nejm.org August 30, 2018
Nystagmus; Wernicke Encephalopathy	*N Engl J Med* 377 e5 July 27, 2017
Osteomalacia	*N Engl J Med* 370; 6 nejm.org February 6, 2014
Rickets	*N Engl J Med* 369; 9 nejm.org August 29, 2013
Schizophrenia	*N Engl J Med* 381; 18 nejm.org October 31, 2019
Scurvy	*N Engl J Med* 374; 14 nejm.org April 7, 2016
Scurvy	*N Engl J Med* 382; 20 nejm.org May 14, 2020

Form 10. Combined Patient History Form

Name _____   Date _____

Part 1. Nutrition History: Food, Caffeine, Alcohol, Fluid, Weight

Nutrients and nutritional status may affect a person's mental health. Please share the following general information about your food intake and eating habits to help assess whether or not you might benefit from further assessment:

(Check, Circle, or Write in Specific Foods)

I. What is the first time you eat during a typical week day? _____

On weekends? _____

What do you commonly eat/drink at that time?

*Coffee/Tea Juice Fruit Milk Toast/Muffin/Bagel/Biscuit Cereal Waffle / Pancake

*Egg/ham/sausage/cheese   Bacon/butter/margarine/peanut butter Yogurt

Smoothie   Protein Shake   Popcorn   Other

_____

II. Do you eat a meal or snack around midday (11AM–2 PM)
Yes / No
What do you commonly eat/drink at that time?

*Milk / Shake or malt / Juice/ Coffee/ Tea / Soft Drink / Beer/ Wine / Water

*Hot or cold sandwich   Chips Fries Soup Salad Pickles

*Meat / Poultry / Fish / Bean / Cheese main dish   Pizza

*Potato Rice Pasta   Bread / Tortilla / Roll / Crackers / Pita / Bagel /

*Vegetables – Raw / Cooked / Juice   Fruit – Whole / Juice

*Dessert – Ice cream / Cookies / Cake / Pie / Candy / Pudding / Jello /

*Other _____

III. Do you eat a meal or snack between 3:00 and 8:00 PM? Yes / No

What do you commonly eat/drink at that time?

*Milk / Shake or malt / Juice/ Coffee/ Tea / Soft Drink / Beer/ Wine / Water

*Hot or cold sandwich   Chips   Fries   Soup   Salad

*Beef / Pork / Poultry / Fish / Bean   main dish   Pizza

*Potato   Rice   Pasta   Bread / Tortilla / Roll / Crackers / Pita / Bagel /

*Vegetables – Raw / Cooked / Juice   Fruit – Whole / Juice

*Dessert – Ice cream / Cookies / Cake / Pie / Candy / Pudding / Jello /

*Other_____   (continued)

Nutrition History: Food, Caffeine, Alcohol, Fluid, Weight

Name_____ Date_____

IV. Do you eat a meal or snack in the mid-to-late evening? Yes / No Meal / Snack

What do you commonly eat/drink at that time? (Circle or name specific other foods)

Chips   Sweets / Deserts / Candy   Left-overs   Sandwiches
Cheese   Crackers   Nuts

Beverages: Coffee   Tea   Beer   Wine   Soft drink   Milk or cocoa
Water

Other _____

* I do / don't eat on a regular schedule.

* I do / don't eat very differently on weekdays than on weekends:

What changes? Different time?   Different foods?   Different place?

Other_____

*Some foods that I eat often, but they aren't on the lists above, include:

_____

*Are there any foods you never eat because of allergies, because you dislike them, or any other reason? Yes / No   Which foods?

_____

_____

*Yes / No I often / sometimes have low calorie / low fat or other "diet" foods/beverages such as

_____

* Would you say you eat small, medium, or large servings? (circle one)

*My height is _____ My weight is _____lbs.

I'm OK with this weight Yes No

I recently gained/lost _____lbs.   I would like to weigh _____

*Do you have any additional comments?

_____

_____

Thank you

Part 2. Family History and Allergies

Name_____   Date_____

Do   you   have   any   food   allergies?   No   Yes

_____

If Yes, What foods are you allergic to? _____

_____

Family history: Do you or a relative have any of the following conditions? (Check any that apply)

Condition	Me	A Relative?; Relationship (parent, cousin, etc.)
Anorexia Nervosa or Bulimia or Other Eating Disorder ?		
Alcohol Intolerance or Abuse?		
Celiac Disease or Wheat Intolerance		
Irritable Bowel Syndrome or Other Intestinal Disorder		
Heart Disease		
Kidney Disease		
Hemochromatosis (Iron Accumulation in the Blood)		
Thyroid Disorder		
Goiter (Enlargement of Thyroid Gland)		
Phenylketonuria (PKU)		
Galactosemia		

Part 3. Supplement Intake

Name_____   Date_____

It is important to include nutrition in improving your mental and physical health.

Please share the following:

1. Do you take any nutritional supplements?   No ____   Yes ____

(Please circle the supplements you take)

Vitamins   Vit A   B-1(thiamin)   B-2 (Riboflavin)
           B-3 (niacin   B-6 (pyridoxine)   B-12 (cobalamin)
           C (ascorbic acid)   D (ergosterol),   E (tocopherol)
           K (menaquinone)   Folic acid (B-9)
           Multi-vitamin supplement_____

Minerals   Calcium (Ca)   Iron (Fe)   Magnesium (Mg)
           Zinc (Zn)   Selenium (Se)   Chromium (Cr)
           Multimineral supplement _____
           Do you use iodized salt? Yes No Don't know

Other   Omega-3 fatty acids   Fish oil   Amino acids _____
           Protein powder   Whey   Casein
           Liquid meals: Ensure   Boost   Other _____

Brand names

_____

_____

_____

Part 4. Nutrient : Medication Interactions

Name_____ Date_____

Do you take any of the following medications?

Name (Prescription)	Dose	Regularly How Often?	Occasionally How Often?
Thyroid			
Omeprezole			
Isoniazid			
Lithium			
Methotrexate			
Zantac			
Statins			
Metformin			
MonoAmineOxidase Inhibitor			
Antiepileptics: Dilantin, Valproate, Carbamazepine			

Name (Nonprescription, Over the Counter)	Dose	Regularly How Often?	Occasionally How Often?

I have / don't have digestive problems (upset stomach, nausea, vomiting)

How often?_____

I have / don't have constipation, diarrhea

How often?_____

Form 9. Permission to Photograph

I, _____, grant permission

to _____

to photograph (who, area)_____

_____

for purposes of planning and evaluating nutritional treatment and for educational purposes.

I understand that I will not be personally identified in any professional educational presentations or published collections of these photographs.

_____
Signature   Date

or _____

Signature of Responsible Person; Relationship Date

_____
Signature of Clinician   Date

Follow-up photographs

Date_____

Date_____

Table 7 Food Sources of Selected Nutrients

The list below suggests foods that are sources of various nutrients.

See also **https://ndb.nal.usda.gov/ndb/** or **https://ods.od.nih.gov/factsheets/list-VitaminsMinerals/** or **https://www.nutrition.gov/whats-food**

NUTRIENT	FOODS WITH MEANINGFUL AMOUNTS OF NUTRIENT	NOTES
Carbohydrate		
Starches	Breads, rice, noodles, potatoes, winter squash, corn, peas, beans, fruit, quinoa	
Fiber	Whole-grain foods: oatmeal, brown rice, whole wheat (cereal, bread, pasta, bagels, muffins, waffles, quinoa	
Sugar	Desserts, soft drinks, candy, jelly, syrup, most granola bars, sweetened cereals	Recommended: not > 10% of calories 1,000 cal: 10% = 200 cal/day; 1 tsp sugar = 16 calories
Galactose	Omission of milk from the diet eliminates major source of galactose	Lactose-free intake necessary for infants with specific genetic conditions. Prescription liquid supplement available.
Fats		
Saturated	Butter, beef, pork, shortening, lard, solid margarines, coconut	
Polyunsaturated	Sunflower, corn, soybean, and flaxseed oils, walnuts, flax seed, fish	
Monounsaturated	Olive, avocado, canola, peanut, macadamia, almonds, pecans, hazelnuts	
Omega 3 fatty acids linolenic acid	Flaxseed (ground), oils (canola, flaxseed, soybean), nuts, and other seeds (walnuts, butternuts and sunflower seeds) mackerel, sardines, herring	
Omega 6 fatty acids linoleic acid	Oils of safflower, soybean, sunflower, walnut, and corn	

Proteins		
Animal proteins	Chicken, turkey, beef, pork, lamb, goat	
Dairy proteins	Milk, cheese, yogurt	
Fish proteins	Salmon, trout, tuna, oysters, shrimp, crab, mackerel, sardines, herring, haddock, tilapia, and others	
Vegetable proteins	Beans/Legumes: soy, garbanzo, black, kidney, pinto, navy	
Extracted proteins: whey, casein, soy, etc.	Protein powders, dry skim milk powder, egg protein, proteins from beans and grains, etc.	
Amino acid phenylalanine (Phe)	Animal proteins contain Phe	2–10 mg/dl Phe in blood required for N development; supplements used
Vitamins		
A	Green and orange vegetables: spinach, kale, carrots, acorn squash, broccoli, tomatoes	UL = 3,000 mcg/day
B-1 – Thiamin	Whole-grain or fortified bread, cereals, pasta, brown rice, pork	
B-2 – Riboflavin	Milk, yogurt, cheese	Spinach, almonds, and mushrooms have one-third to one-half as much per average serving
B-3 – Niacin	Proteins	UL = 35 mg/day
B-6 – Pyridoxine	Spinach, brussels sprouts, romaine lettuce, avocado, fortified cereals, pasta, rice, fortified: breads, orange juice	UL = 100 mg/day
B-9 – Folic acid	Meat, eggs, beans, legumes, seafood, fish, fortified: breakfast cereals, breads, orange juice	UL = 1,000 mcg/day
B-12 – Cobalamin		
C – Ascorbic acid	Oranges, grapefruit; blueberries, broccoli, tomatoes	UL = 2,000 mg/day

(Continued)

Table 7 (Continued) Food Sources of Selected Nutrients

The list below suggests foods that are sources of various nutrients.

See also **https://ndb.nal.usda.gov/ndb/** or **https://ods.od.nih.gov/factsheets/list-VitaminsMinerals/** or **https://www.nutrition.gov/whats-food**

NUTRIENT	FOODS WITH MEANINGFUL AMOUNTS OF NUTRIENT	NOTES
Choline	Eggs, liver, and peanuts; major contributors in the U.S. are meat, poultry, fish, dairy foods, and egg-based dishes. A choline metabolite, betaine, is found in spinach, beets, wheat, and shellfish	UL = 3.5 g/day
D – Ergosterol	Sunshine on skin, cod liver oil, fortified milk	UL = 100 mg/day
E – Tocopherol	Sunflower seeds/oils, almonds, hazelnuts, peanut butter	UL = 15 mg (22 IU)/day
Minerals		
Calcium	Milk, cheese, yogurt	UL = 2,500 mg/day
Chromium	Meat and whole-grain products; some fruits, vegetables, and spices are relatively good sources	
Copper	Organ meats, nuts and seeds, chocolate, and shellfish have the highest copper content	UL = 10,000 mcg/day
Iodine	Seaweed (kelp, nori, kombu, and wakame) generally high in iron but highly variable in its content. Other good sources include iodized salt, seafood, dairy products (especially milk), are major contributors of iodine to the American diet. Also present in human breast milk and infant formulas. It is especially important for pregnant women to consume adequate iodine	UL = 1,100 mcg/day Sea salt is not iodized
Iron	Meat, eggs, beans, legumes, seafood, fish, fortified breakfast cereals	UL = 45 mg/day
Magnesium	Green vegetables (spinach, kale, broccoli), nuts	UL = 350 mg in supplement in addition to food intake
Selenium	Many supplements contain selenium; high intake of Brazil nuts	UL = 400 mcg/day
Zinc	Whole grains, meat	UL = 40 mg/day

Table 8 Nutrients, the Brain, CNS, and Mental Status
See Appendix for References for Table 1.

NUTRIENT	BRAIN /CNS PHYSICAL STATUS	S/S MENTAL STATUS	DRI / DOSES	FOOD FACTORS
Amino acids (AA)				
9 of the 20 AA are essential	Synthesis of neurotransmitters (serotonin, dopamine, norepinephrine, acetylcholine, histamine, and glycine) is influenced by the AA in the diet			
Phenylalanine (Phe) An essential amino acid	Necessary for enzyme phenylalanine hydroxylase; Single Polymorphism on Chromosome 4	If untreated; Abnormal development from birth; Low IQ;	2-10 mg/dl Phenylalanine in blood required for N development	Animal proteins contain Phe and are restricted; AA supplement required.
Tryptophan (Trp) An essential amino acid	Precursor of neurotransmitter Serotonin Tryptophan is converted to 5-HTP, which is used to synthesize serotonin. Serotonin turnover rate in the central nervous system may be modulated by DHA.	Serotonin may play a role in depression, insomnia, weight loss, migraine headaches, fibromyalgia, and anxiety. Normal levels of serotonin lead to better impulse control, decreased irritability, less feelings of depression, feeling relaxed, and more able to go to sleep, less cravings for sweets and decreased tendency toward aggression.		Found in foods with protein. A small amount of protein (4% of calories) in a high-carbohydrate meal is sufficient to block any meal-induced increases in the ratio of plasma tryptophan to other AA.

(Continued)

Table 8 (Continued) Nutrients, the Brain, CNS, and Mental Status
See Appendix for References for Table 1.

NUTRIENT	BRAIN/CNS PHYSICAL STATUS	S/S MENTAL STATUS	DRI/ DOSES	FOOD FACTORS
Tyrosine A conditionally essential amino acid	Becomes essential if phenylalanine is not adequate Tyrosine is the precursor of the catecholamines; alterations in the availability of L-tyrosine to the brain can influence the synthesis of both dopamine and norepinephrine.	Symptoms of serotonin syndrome (excesses) include confusion, agitation, anxiety, hypomania, or in severe cases, coma. Autonomic symptoms may include profuse sweating and hyperthermia and/or muscle rigidity Reported by military to reduce cognitive signs and symptoms of physical stress such as extreme cold, hypoxia and flight conditions.	Tyrosine can spare 78% of the dietary phenylalanine need. Optimal proportions of dietary phenylalanine and tyrosine have been shown to be 60:40, respectively.	Found in protein foods.

Carbohydrates			
Glucose is the preferred fuel for the brain. If insufficient: Switches to fatty acids as energy source. Hypoglycemia impairs simple brain functions related to task-specific localized reductions in brain activation. For tasks with greater cognitive load, the brain recruits other brain regions. Alters insulin–tryptophan–serotonin pathway.	Symptoms of hypoglycemia have been classified as (1) autonomic (palpitations, sweating, shaking, and hunger), (2) neuroglycopenic (confusion, drowsiness, inability to concentrate, speech difficulty, and blurred vision). and (3) nonspecific (nausea and headache). S/S: fatigue, compulsive eating, Food cravings, anxiety, nervousness, confusion with extremely low or excessive glucose levels. Hyperglycemia: Individuals may be unable to concentrate, unable to pay attention, and may be unusually moody and irritable; they may be classified as being (1) alert, (2) drowsy, (3) in a stupor, or (4) in a coma. Contrary to popular belief, when observed following a high-sugar drink, children's behavior was slowed down. Normal children's behavior was slowed down in ~ three hours; children diagnosed as psychiatrically disturbed slowed down in one hour.	Hypoglycemia: blood glucose level below 70 mg/dl. Blood glucose level above 250 mg/dl has been observed to change mental status. In health, autonomic activation occurs at plasma glucose of 3.0–3.6 mmol/L. Reaction times become slower at about 3 mmol/L; short-term memory deteriorates at 2.5 mmol/L.	Sugars enter the bloodstream within a few minutes. Starches that contain fiber enter the blood more slowly.

(Continued)

Table 8 (Continued) Nutrients, the Brain, CNS, and Mental Status

See Appendix for References for Table 1.

NUTRIENT	BRAIN/CNS PHYSICAL STATUS	S/S MENTAL STATUS	DRI / DOSES	FOOD FACTORS
Galactose	Galactose: the simple sugar derived from lactose in milk. Galactosemia- SNP- single polymorphism; Chromosome 9 Lack of galactose-1-phosphate uridyl transferase (GALT); possible lack of galactokinase/UDP4-epimerase.	Functions of galactose: cell membranes, glycolipids, glycoproteins, sulfation pathways and phenol degradation. SNP results in behavioral development that does not follow normal sequence Low IQ; abnormal development from birth		Omission of milk from the diet eliminates major source of galactose. Lactose-free intake until 1 year old Rx supplement beverage
Lipids includes *triglycerides, *EFA and others (such as N-9 Oleic acid), *sterols (such as cholesterol and hormones), *phospholipids	EFAs are necessary for building and maintaining the phospholipid layer that surrounds nerve axons (the myelin sheath; crucial for nerve transmission between the brain and the rest of the body). EFAs include the following: Linoleic acid: 0-6 fatty acid with 18 C, two double bonds Linolenic acid: 0-3 fatty acid with18 C, three double bonds	Linoleic >> Gamma linoleic acid >> Arachidonic Linolenic >>EPA >> DHA High 0-6: 0-3 deficiency linked to depression Used as an adjunct to other therapies Ratio of 0-6 to 0-3 influences neuro-transmission of serotonin and catecholamine Ratio of 0-6:0-3 in the brain and central nervous system is affected by dietary intake. Levels and ratios of EFA appear to be linked to anger, violence, hostility and aggressive behavior.	DRI: **Linoleic acid** (0-6): 5%–10% of calories; 14–17 grams/day for men and 11–12 grams/day for women. **Linolenic acid** (0-3): 0.6%–1.2% of calories; 1.6 grams/day for men; 1.1 grams/day for women The National Institutes of Health recommend foods that provide 650 mg per day of 0-3 fatty acids.	Caution regarding use of fish oil that may provide excessive vitamin A or mercury (Hg) Caution regarding 0-3 supplements use simultaneously with blood thinners such as warfarin or aspirin. Intake ratio of 3:1 EPA:DHA decreases the overactivity of a phospholipase enzyme that breaks down phospholipid structures.

				Food Sources linoleic
Lipids includes (Continued)	Arachidonic acid: 20 C: three unsaturated bonds. Conditionally essential depending on amount of linoleic and linolenic EFA. EPA (Eicosapentaenoic acid) 22 C: O-3 fatty acid; plays a role in membrane function. Can be synthesized from alpha linolenic acid (ALA), although not efficiently (< 5%). DHA (DocosaHexaenoic Acid) 20 C; O-3 fatty acid; plays a role in membrane structure. DHA is present in large amounts in neuron membrane phospholipids, a structural component of the human brain, cerebral cortex, skin, sperm, testicles, and retina. EPA and DHA are essential for proper fetal development and healthy aging. Found in abundance in the brain and retina.	DHA stabilized mood in bipolar disorder. O-3 levels shown low in schizophrenia. The actions of EPA and DHA resemble the actions of two psychotropic drugs: lithium and valproate. Cholesterol levels below 160mg/dL linked to depression and suicide risk. Cholesterol may be a biomarker for EFA. 1-gram/day of EPA provided benefit for depression, anxiety, sleep, lassitude, libido, and suicidality. EFA may be deficient in individuals with anorexia nervosa. Schizophrenia, bipolar disorder, and severe depression linked to two genes. One is common to all three; the other influences which symptoms occur. Results in removal of EPA, DHA, AA, and GLA from phospholipid structure.	The American Heart Association advocates 1 gram (1,000 mg) of O-3 in enriched foods daily. At an Institute of Nutrition symposium, Joseph R. Hibbeln commented 180–500 mg/day EPA is inadequate, 750 mg/day possibly adequate, and 1,000 mg/day of EPA appears to be effective in reducing psychotic, depressive, and aggressive symptoms. The estimated intake in the United States of EPA and DHA is 180 mg/day.	acid (O-6): Oils from corn, sunflower, safflower, soybean, cottonseed, poultry fat, nuts, and seeds. **Food sources of linolenic** acid (Omega-3): Oils from flaxseed, canola, walnut, wheat germ, and soybean; nuts and seeds such as flaxseeds, walnuts, soybean kernels, and butternuts. EPA and DHA may be obtained from pacific oysters, mackerel, salmon, bluefish, mullet, sablefish, menhaden, anchovy, herring, lake trout, sardines, and tuna.

(Continued)

Table 8 (Continued) Nutrients, the Brain, CNS, and Mental Status
See Appendix for References for Table 1.

NUTRIENT	BRAIN /CNS PHYSICAL STATUS	S/S MENTAL STATUS	DRI / DOSES	FOOD FACTORS
	EFA intake linked to postpartum depression. Joseph R. Hibbeln et al. showed that the relationship of DHA to CSF (cerebral spinal fluid). 5-HIAA(5-hydroxy-indoleacetic acid) and HVA (homovanillic acid) were significantly different in alcoholic groups. (5-HIAA and homovanillic acid HVA are metabolites of neurotransmitters.) Studies to determine if supplements of EFA can influence neurotransmitters and modify impulsive behaviors were suggested. Links to Alzheimer's disease, dementia, and loss of cognitive ability have been questioned. Fiala (2015) study showed improved cognition with O-3 FA suppl.			

Minerals

Minerals – Toxic		
Tolerable upper levels (ULs) have been defined for the following minerals: boron (B), copper (Cu), fluoride (F), iodine (I), manganese (Mn), molybdenum (Mb), nickel (Ni), selenium (Se), vanadium (V), and zinc (Zn). These minerals are required, but excess intakes are known to have detrimental effects.	Toxic minerals most often associated with foods are mercury and lead. Methyl mercury consumed is taken up within one to two days; in greater concentration by the brain than other tissues. S/S: fatigue, headache, decrease in memory and concentration. Hightower: Reports it took 41 weeks to reach normal blood levels by those with the highest excess intakes Pottery and lead-glazed ceramics may have high levels of lead. Ayurvedic remedies were found by the Centers for Disease Control and Prevention to have high levels of lead, arsenic, and mercury.	Fish highest in mercury: Mackerel (king) marlin orange roughy shark wordfish tilefish tuna (bigeye, ahi).

(Continued)

Table 8   (Continued) Nutrients, the Brain, CNS, and Mental Status
See Appendix for References for Table 1.

NUTRIENT	BRAIN/CNS PHYSICAL STATUS	S/S MENTAL STATUS	DRI / DOSES	FOOD FACTORS
Chromium (Chr)	Chromium is involved in the metabolism of carbohydrate; chromium supplements have been tentatively shown to reduce carbohydrate cravings. Weight gain, a side effect of several psychotropic medications, is a common cause of discontinuation of prescribed medications.	Atypical depression, a common form of major depression, is characterized by a distinct combination of symptoms including mood swings, carbohydrate cravings, weight gain, rejection sensitivity, and lethargy.	AI 20–30 mcg/day for adults	Meat and whole-grain products, as well as some fruits, vegetables, and spices, are relatively good sources. Food content is affected by agricultural and manufacturing processes.
Copper	Wilson's disease is an autosomal recessive disorder of copper storage. Copper accumulates in the liver and neuronal tissue when the gene ATP7B is defective. The Wilson protein is required for copper to be removed from tissue. Use of a chelating agent, along with avoidance of foods high in copper, reduces copper stores.	Dementia resulting from copper accumulation is treatable with chelating agents.	D-penicillamine, a chelating agent is an antimetabolite of pyridoxine (vitamin B_6); a supplement of B_6 of 12.5–25 mg or 75–150 nanomoles/day is recommended.	Food sources: organ meats, nuts and seeds, chocolate, and shellfish have the highest copper content.

| Iodine | Lithium may induce goiter; incidence ranges from 20% in patients residing in iodine-replete areas to 87% in patients residing in or emigrating from iodine-deficient areas or who are on long-term lithium therapy. Iodine deficiency during pregnancy, leads to impairment in brain development, resulting in permanent neurological damage of the fetus. At 10–18 weeks gestational age, iodine deficiency may result in mental deficiency, deafness, mutism, motor disorders, and basal ganglia calcification (cretinism). | The American Academy of Pediatrics issued a statement on iodine, saying about one-third of pregnant women in the United States are iodine-deficient (May 26, online/ June print, 2014; *Pediatrics*). Even marginal iodine deficiency can decrease brain functioning, the report said. Hypothyroidism: fatigue, hypersomnolence, cognitive impairment (forgetfulness), difficulty concentrating or learning. Hyperthyroid: fatigue, insomnia, mood instability, irritability, anxiety, nervousness, decrease or increase in appetite. | DRI 150 mcg/day for adults WHO recommends 250 mcg/ day during pregnancy Pregnant and breast-feeding women should take a supplement that includes at least 150 micrograms of iodide, and use of iodized table salt; combined intake from food and supplements should be 290 to 1,100 micrograms a day. Potassium iodide is the preferred form | Seaweed (kelp, nori, kombu, and wakame) generally high in iodine but highly variable in its content. Other good sources include seafood and dairy products, especially milk, a major contributor of iodine to the American diet. Also present in human breast milk and infant formulas. |

(*Continued*)

Table 8   (Continued) Nutrients, the Brain, CNS, and Mental Status
See Appendix for References for Table 1.

NUTRIENT	BRAIN/CNS PHYSICAL STATUS	S/S MENTAL STATUS	DRI / DOSES	FOOD FACTORS
Iron (Fe)	Iron– an essential nutrient, but it may also be a potent toxin. Functions that are specific to neurological activity include synthesis of dopamine, serotonin, catecholamines, and possibly GABA and myelin formation. The highest levels of iron in the brain occur in the basal ganglia. White matter throughout the brain, largely in the myelin sheath, is a major site of iron concentration. Iron uptake is maximal during rapid brain growth.	In older children, a hemoglobin of less than 9g/dl iron resulted in decreased attentiveness, narrow attention spans, perceptual restriction, and poor performance on the Bayley Mental Index. Iron is a significant component of senile plaques and iron encrustation of the brain's blood vessels and is found in the blood vessels of senile patients. Also observed are elevated levels of iron in other parts of the brain: the hippocampus, the amygdala, the basal nucleus of Maynert, and the cerebral cortex.	DRI Adults: Females 31–50 years: 18 mg/day Males and other females: 8 mg/day TUIL: 45 mg/day Recommended intake assumes 75% of iron is from heme iron sources Requirement for individuals consuming a vegetarian diet is approximately twofold greater than for those consuming a nonvegetarian diet.	Food sources Fortified cereals, oysters, legumes, dark chocolate, tofu.

| Magnesium (Mg) | Magnesium is involved in over three hundred metabolic reactions. Serum magnesium may not reflect body stores. Depression of the central nervous system may occur in the range of serum levels of 5–10 mEq/L. N serum Mg = 1.5–2.5 mEq/L. Involved in active transport of calcium and potassium ions across cell membranes, important to nerve impulse conduction Birmingham et al found that 60% of patients with anorexia nervosa had low serum magnesium at admission and up to three weeks into the refeeding program. | Symptoms of magnesium depletion may include depression and psychosis, as well as seizures and numerous physical signs. Cellular targets for lithium action involve magnesium-activated enzymes, which are inhibited by lithium. Lithium inhibits inositol monophosphate, which decreases ability of neurons to generate intermediate compounds. Depression of the central nervous system may occur in the range of serum levels of 5–10 mEq/L. Normal serum magnesium range is 1.5–2.5 mEq/L. Excess magnesium may exhibit with confusion and lethargy in addition to physical symptoms of toxicity. | Adults: 300–420 mg/day Children: 80–130 mg/day (dependent on age) | Food sources: green leafy vegetables, legumes, nuts, seeds, whole grains, and cereal, other fortified foods, dark chocolate. |

(Continued)

Table 8   (Continued) Nutrients, the Brain, CNS, and Mental Status
See Appendix for References for Table 1.

NUTRIENT	BRAIN /CNS PHYSICAL STATUS	S/S MENTAL STATUS	DRI / DOSES	FOOD FACTORS
Selenium (Se)	Greater uptake of selenium in the brain than other tissues. At times of deficiency, the brain retains selenium to a greater extent. Eleven selenoproteins have been identified. Several act as antioxidants, are interdependent with vitamin E, iron, zinc, and copper. The function of selenium in the brain is related to functions of selenium–dependent enzymes (a) glutathione peroxidase (Se-GPX) and (b) a selenocysteine-containing enzyme, Type I thyroxine deiodinase. Se-GPX activity occurs in myelin tissue. Se-GPX protects polyunsaturated fatty acids (PUFA) from peroxidation.	Toxicity (selenosis): fatigue, irritability, depression, memory loss, hair loss, diarrhea, abdominal pain Neurologic symptoms of toxicity may include tremor, muscle spasms, restlessness, confusion, delirium, and coma. Other symptoms include fatigue, irritability, and depression. Gradual nervous system toxicity from environmental contamination with residues of pesticides may result in S/S of learning, memory loss, hyperactivity, aggressiveness, and defensiveness and may occur at low doses. Acute toxicity can cause death in hours or days. Neither chelation nor emesis is recommended. Deficiency: memory loss, hyperactivity, aggressiveness, and defensiveness. Patients with schizophrenia treated with clozapine may have low selenium.	RDA/DRI Women: 55mcg/day Men: 70 mcg/day One study showed deficiency in ≈50% of both sexes. Children/adolescents: 30–40 mcg/day older	Food sources: Brazil nuts provide 68–91 mcg per nut. Many whole grains and dairy products, including milk and yogurt. Some ready-to-eat breakfast cereals are fortified with selenium. Pork, beef, turkey, chicken, fish, shellfish, and eggs contain high amounts of selenium. Selenium in soil is reflected in selenium levels in grains grown in the soil.

(Continued)

| Selenium (Se) (Continued) | Selenium deficiency could exacerbate iodine deficiency. Average serum selenium in the United States is ~1.27–1.78 μmol. In serum or plasma: Acute toxicity: 400–30,000 μg/L Chronic toxicity: 500–1400 μg/L Not toxic: <1,400 μg/L Urine assessment: selenium should be < 100 mcg/L. Strong garlic-like odor is commonly present in both acute and chronic poisoning, attributed to the volatile metabolite dimethylselenide. Selenocysteine is the only trace element to be incorporated into the genetic code. Selenium has been shown to regulate gene expression. Spermatozoa have the highest concentration of selenium in the mammalian body; deficiency may impair the synthesis and motility of sperm. | SELENBP1 gene expression strongly positively correlated with presence of psychosis across diagnoses of schizophrenia and bipolar disorder. |

Table 8 (Continued) Nutrients, the Brain, CNS, and Mental Status

See Appendix for References for Table 1.

NUTRIENT	BRAIN/CNS PHYSICAL STATUS	S/S MENTAL STATUS	DRI/DOSES	FOOD FACTORS
Sodium — Na	When amount of sodium in fluids outside cells drops, water moves into the cells to balance the levels. This causes the cells to swell with too much water. Brain cells are especially sensitive to swelling, and this causes many of the symptoms of hyponatremia. Hyponatremia can be caused by:Overhydrating during exercise Burns that affect a large area of the body Diarrhea, diuretic medicines, which increase urine output, heart failure, kidney disease, liver cirrhosis, syndrome of inappropriate antidiuretic hormone secretion (SIADH) sweating, vomiting	Symptoms of hyponatremia include confusion, convulsions, fatigue, headache, irritability, loss of appetite, muscle spasms or cramps, muscle weakness, nausea, restlessness, vomiting Acute hyponatremia, which occurs in less than 48 hours, is more dangerous than slowly developing hyponatremia. Hyponatremia: serum Na < 135 mEq/LSevere: serum level below 125 mEq/L. Serum Na < 115 meq/L: intracerebral osmotic fluid shifts and brain edema. Polydipsia in psychiatric patients may be the cause of hyponatremia; possibly a side effect of typical antipsychotics.	Free water restriction (< 1 L/day) is generally the treatment of choice. Hypertonic solution only for overtly symptomatic hyponatremia (e.g., seizures, severe neurologic deficits), hypertonic (3%) saline should be used.	Iodized salt is often the main source of Iodine in the diet. Some preservatives may contain iodine.

| Zinc | Zinc participates in the activity of over 200 enzymes. Zinc is an important cofactor for metabolism of neurotransmitters, fatty acids, prostaglandins, and melatonin, and it indirectly influences dopamine metabolism. Zinc deficiency leads to both primary and secondary alterations in brain development and growth. Zinc is utilized for incorporation of thymine into DNA, one of the four bases forming the structure of DNA. Inadequate zinc produces alteration in gene transcription. One of the most striking alterations is primary neural tube defect. | Low serum zinc has been reported in patients with treatment-resistant depression. The severity of depression was correlated with the decrease in zinc levels. Six weeks of zinc supplementation improved response to antidepressant therapy by over 50% and was sustained over 12 weeks of the study. Investigators hypothesize that zinc may influence 5-HT uptake in the cell body, block reuptake at the synapse, or increase synaptic release; Significantly higher copper and lower zinc concentrations were found in patients who experienced delirium tremens or prolonged hallucinatory states during withdrawal from alcohol. | DRI Adults: 8-11 mg/day TUIL 40 mg/day Absorption is lower when consuming vegetarian diets than for nonvegetarian diets. For those consuming a vegetarian diet recommendation is approximately twofold greater than for those consuming a nonvegetarian diet. | Food sources: oysters, red meat (e.g., beef, veal, pork, and lamb), and fortified cereals. |

(Continued)

Table 8 (Continued) Nutrients, the Brain, CNS, and Mental Status
See Appendix for References for Table 1.

NUTRIENT	BRAIN /CNS PHYSICAL STATUS	S/S MENTAL STATUS	DRI / DOSES	FOOD FACTORS
	Zinc and folate both influence genetic transcription and play a part in development of the central nervous system during pregnancy. Zinc is more abundant in gray matter of the brain than white matter. Concentration of zinc in the hippocampus and amygdala are assumed to be necessary for experiential learning. Highest levels of zinc are observed in the hippocampus. Deprivation effects are seen most strongly in poor growth of cerebellum and Purkinje cells.			

Zinc
(Continued)

Indirect effects of zinc
deficiency may be observed
in the lipid content of the
developing brain. This
may suggest alterations of
function in the brain due
to changes in membrane
characteristics.
Changes in brain zinc
concentration are observed
in Alzheimer's disease, Down
syndrome, epilepsy, multiple
sclerosis, retinal dystrophy,
and schizophrenia.

Vitamins

Vitamin deficiency diseases
involve both the central
nervous system and the
peripheral nervous system,
an attribute they share
with few other categories of
disease.

(Continued)

Table 8 (Continued) Nutrients, the Brain, CNS, and Mental Status
See Appendix for References for Table 1.

NUTRIENT	BRAIN/CNS PHYSICAL STATUS	S/S MENTAL STATUS	DRI/ DOSES	FOOD FACTORS
Vitamin A (retinol)	Chronic hypervitaminosis A can give rise to increased intracranial pressure.	Acute toxicity of vitamin A causes symptoms of headache, dizziness, irritability, and drowsiness.	The DRI and (AI) for vitamin A is 700–900 µg/day. Tolerable UL for adults is 3,000 µg per day. The safe level for children is less than half of this, depending on age.	Food sources of pre-formed vitamin A is found in animal sources, including dairy products, fish, and meat (especially liver). Most dietary pro-vitamin A comes from leafy green vegetables, orange and yellow vegetables, tomato products, and fruits.
B-1 – Thiamin	Deficiency: Beri-beri, Wernicke-Korsakoff psychosis Thiamin: a cofactor for three enzymes: transketolase, pyruvate dehydrogenase, and alpha-ketoglutarate dehydrogenase. B-1-dependent enzymes required for glucose metabolism; Required for metabolism of branched chain AA Activity of erythrocyte transketolase enzyme	Fatigue Confusion Wernicke-Korsakoff psychosis: Disoriented to time and place, indifference to surroundings, inability to grasp meaning, irrational remarks Severe hyperemesis gravidarum can develop into Wernicke-Korsakoff psychosis Deficiency with foot drop and other S/S of B-1 deficiency reported following bariatric surgery	DRI: 1.2 to 1.4 mg/day Recovering alcoholics or other deficient states: First 100 mg B-1 IV, Then 100 mg IM 5days Then maintenance of 1–1.5 mg/d oral supplement with adequate meals	Food sources: enriched, fortified, or whole-grain products, mixed foods whose main ingredient is grain, ready-to-eat cereals. Pork Legumes

B-1 – Thiamin (Continued)	Thiamin pyrophosphate activity (TPP) Thiamin functional load test % of increased activity: 0%–15% increase = Normal 16%–20% increase = Low 20% increase = Deficient Deficiency: Brain lesions, destruction of nerve cells and fibers Nutriogenomic Risk factor For alcoholism ADHS*1 allele ADH2*1/2*1 genetoype Thiaminases in raw fish, clams, shrimp, and mussels inactivate the thiamin present.	Subjects who took thiamin supplement more than doubled their scores on clear-headedness and mood and increased their quickness of reaction time.

(Continued)

Table 8 (Continued) Nutrients, the Brain, CNS, and Mental Status

See Appendix for References for Table 1.

NUTRIENT	BRAIN/CNS PHYSICAL STATUS	S/S MENTAL STATUS	DRI/ DOSES	FOOD FACTORS
B-2 – Riboflavin	Riboflavin is essential to enzymes such as monoamine oxidase. Monoamine oxidase inhibitors (MAOI) are one type of antidepressant medication. Phenothiazines and tricyclic antidepressants (two types of psychotropic drugs) competitively inhibit flavin coenzyme biosynthesis, but alone they are not believed to induce deficiency. Riboflavin: essential in energy metabolism as flavin adenine dinucleotide (FAD) and flavin mononucleotide (FMN). Thyroid hormones and adrenal steroids enhance FMN and FAD synthesis.	Haller: Severe riboflavin deficiency resulted in changes of scores on the hypochondriasis, depression, hysteria, psychopathic–deviate, and hypomania scales on the Minnesota Multiphasic Personality Inventory (MMPI) assessment instrument	Men 14 years or older: 1.3 mg; Women 14–18 years, 1 mg; Women over 18 years: 1.1 Children 9–13 years: 0.9 mg; Infants – 8 years : 0.3 mg–0.06 mg	Food sources: milk, meat, eggs, nuts, enriched flour, and green vegetables.

B-3 – Niacin, nicotinamide, nicotinic acid	Def disease: Pellagra: S/S dementia, diarrhea, dermatitis		DRI:	Food sources: meats and alternatives are the richest sources followed by fortified enriched grains.
	Pellagra may result from an inadequate diet, alcoholism, and anorexia	In pellagra, fatigue, insomnia, and apathy may precede the development of confusion, disorientation, hallucination, loss of memory, and, eventually, organic psychosis.	14 mg for women 16 mg for men	
	B-3 necessary for energy production – TCA cycle	The mental changes may be due to diminished conversion of tryptophan to serotonin. Pellagra is progressive over several years.	Ten mg/day of niacin accompanied by adequate amounts of dietary tryptophan is sufficient to cure pellagra.	
	Pellagra is basically an encephalopathy, although involvement of spinal cord and peripheral nerves may occur. Early mental symptoms include insomnia, fatigue, anxiety, nervousness, irritability, and depression. The nutritional deficiency may be mistaken for a psychiatric disorder. As the disease advances, slowing and inefficiency of mental processes and impairment of memory become apparent.	Morris: residents who consumed the least niacin (median of 14 mg/day) were 70% more likely to develop Alzheimer's disease than those who consumed the highest amounts. The group with highest intake had a total of 45 mg/day in a combination of diet and supplements. Benefits were observed with the median intake of 17 mg/day.	Slow-release preparations of niacin (given in the form of nicotinic acid) cause less discomfort from flushing; however, acute onset of hepatic toxicity (in two days to seven weeks) can occur with doses of 1,500 mg/day or more.	

(Continued)

Table 8 (Continued) Nutrients, the Brain, CNS, and Mental Status
See Appendix for References for Table 1.

NUTRIENT	BRAIN /CNS PHYSICAL STATUS	S/S MENTAL STATUS	DRI / DOSES	FOOD FACTORS
B-6 – Pyridoxine	Vitamin B_6 plays a role in neuronal excitability, possibly as a result of transsulfuration reactions or in γ-aminobutyric acid (GABA) metabolism. GABA is a physiologic inhibitor of neuro-transmission. Pyridoxine is a cofactor for 5-hydroxy-tryptophan decarboxylase (an enzyme in the biosynthesis pathway of serotonin).	Vitamin B6 plays a role in cognitive development through the biosynthesis of neurotransmitters and in maintaining normal levels of homocysteine in the blood. Chronic pyridoxine intoxication causes sensory polyneuropathy similar to that caused by its deficiency. Excessive amounts of pyridoxine appear to cause degeneration of dorsal root ganglia. Ingestion of 2 g/day (2,000 mg/day) may cause: progressive ataxia, impairment of position and vibration sense, and loss of deep tendon reflexes; reversible with discontinuation of high doses of vitamin B_6.	The DRI 1.3–2.0 mg/day for adults 0.1–0.6 for children up to 8 years	Food sources: fish, beef liver, and other organ meats; potatoes and other starchy vegetables; and fruit (other than citrus). About 75% of vitamin B6 from a mixed diet is bioavailable.

B-9 – Folic acid/folacin				
Forms of vitamin B-9: >Folate – in natural foods >Folic acid –synthetic, in supplements >Folacin – generic term for above >Folinic acid – a derivative, full activity >L 5 tetra Methyl-hydrofolate OR >5 formyl tetrahydrofolate – a natural form found in supplements	Folate is the active form Supplies methyl groups for numerous biochemical reactions Helps repair frequent, small damages to your DNA. Folate is absorbed 85% as efficiently as folic acid; may be related to genetic alterations found in 20%–55% of the population. One genetic alteration is that of the enzyme *5, 10-methylene tetrahydrofolate reductase* (MTHFR), known as the "*MTHFR* c.677C>T" polymorphism; 10% of people of northern European descent and 15% of those with southern Europeans ancestors are estimated to have this genetic substitution	Low folate has been associated with changes in cognition such as memory, attention, and language. Def in pregnancy causally associated with neural tube defects in newborns. One case: After 3 weeks on a low (5 mcg) folate diet, serum folate fell from 7 ngs/ml to below 3 ngs/ml. Onset of anemia and red cell macrocytosis, as well as the mental changes that occurred in the 18th week. The symptoms abated following oral folate replacement of 250 µg. In outpatients with schizophrenia, folate concentrations were inversely related to the Scale for Assessment of Negative Symptoms. Negative symptoms associated with schizophrenia are social isolation, lack of initiative, socially awkward behavior, and discomfort when interacting with people.	The DRI for adults is 400 mcg/day. The tolerable upper intake of folic acid for healthy adults is 1,000 mcg/day. Since 1994 when supplementation of grain foods with folate began, the incidence of neural tube defects has decreased. The program increased mean folic acid intakes in the United States by about 190 mcg/day.	Food sources: Highest: spinach, liver, yeast, asparagus, and brussels sprouts. Good: dark green leafy vegetables, fruits and fruit juices (especially orange juice), nuts, beans, peas, dairy products, poultry and meat, eggs, seafood, and grains.

(Continued)

Table 8 (Continued) Nutrients, the Brain, CNS, and Mental Status
See Appendix for References for Table 1.

NUTRIENT	BRAIN /CNS PHYSICAL STATUS	S/S MENTAL STATUS	DRI / DOSES	FOOD FACTORS
	Impaired homocysteine metabolism r/t this polymorphism higher in depression			
	Total plasma homocysteine is a more sensitive measure of folate status.			
	Children with autism have been shown to have significantly lower baseline plasma concentrations of methionine, SAM (S-adenosylmethionine), homocysteine, cystathionine, cysteine, and total glutathione and higher concentrations of SAH (S-adenosylhomo-cysteine) and oxidized glutathione are consistent with impaired capacity for methylation.			

| B-12 – Cobalamin | Necessary for maintenance of myelin sheath surrounding nerves and communication within the central nervous system. Deficiency: diffuse, uneven, deterioration/degeneration of white matter of spinal cord and sometimes the brain. Physical symptoms of a vitamin B_{12} deficiency include being easily fatigued, being increasingly unsteady on the feet, diminished sensation in the arms or legs, feeling "foggy" or depressed, suffering nerve damage, and a characteristic type of anemia. | Mental signs of vitamin B_{12} deficiency include irritability, apathy, somnolence, suspiciousness, and emotional instability. Symptoms also include memory problems, slowed thinking, mood swings, and trouble concentrating. In depression: decrease in serum/red blood cell folate, serum vitamin B12 and an increase in plasma homocysteine B_{12} appears to be independently related to depression. Memory clinic patients with low B_{12} were treated with B_{12} supplement. Cognitive impairment improved significantly in verbal fluency. There was no significant improvement in dementia patients after treatment. Zucker: symptoms of B_{12} deficiency in psychiatric patients as organic brain syndrome, paranoia, violence, and depression. | DRI 0–6 mo: 0.4 mcg/day 7–12 mo: 0.5 mcg 1–3 yr: 0.9 mcg 4–8 yr: 1.2 mcg 9–13 yr: 1.8 mcg 14–adult: 2.4 mcg An estimated 10%–20% of people in their late 60s or older have some degree of B_{12} deficiency. Acute treatment for pernicious anemia is 1,000 µg intramuscular injections daily followed by weekly injections for a month, then monthly injections for the remainder of the individual's life. | Vitamin B12 is found in animal products, including fish, meat, poultry, eggs, milk, and milk products. Vitamin B12 is generally not present in plant foods. Fortified breakfast cereals are a source of vitamin B12 with high bioavailability. Some nutritional yeast products also contain vitamin B12. |

(Continued)

Table 8 (Continued) Nutrients, the Brain, CNS, and Mental Status
See Appendix for References for Table 1.

NUTRIENT	BRAIN/CNS PHYSICAL STATUS	S/S MENTAL STATUS	DRI/ DOSES	FOOD FACTORS
	Serious problems are neurologic, such as diminished sensations in the foot or inability to walk normally. Optic neuropathy reported with all forms of B_{12} deficiency. Neuropsychiatric disorders without anemia may found in patients with vitamin B_{12} deficiency. Cause appears to be related to the adequacy of Methionine. Infants: Symptoms of a deficiency likely to occur within three to six months after birth to mothers with low B-12 and in six to ten weeks in infants born to and nursed by mothers with pernicious anemia. Possible consequences: abnormal neurological development and mental development Anti-acids & proton-pump inhibitors may induce B-12 deficiency	Nonanemic vitamin B_{12} deficiency can be treated only through an increased supply of methionine or an accelerated production of methionine from homocysteine, a reaction that requires vitamin B_{12}. An inadequate amount of methionine caused by a deficiency of vitamin B_{12} decreases the availability of SAM. SAM is required for important methylation reactions (such as choline formation), that are essential to the maintenance of myelin and prevention of neuropathies. Many adolescents who consumed macrobiotic (vegan-type) diets up to about age 6 and then consumed lactovegetarian or omnivorous diets were deficient in MMA (a measure of vitamin B_{12} status). Psychological tests showed a significant correlation between vitamin B_{12} deficiency and fluid intelligence. Fluid intelligence involves reasoning, the capacity to solve complex problems, the ability to learn, and abstract thinking ability.	Low-dose supplementation (2–37.5 µg/day) of oral cobalamin associated with higher serum levels of cobalamin and improved or normalized cobalamin function. No toxicity or benefit has been recorded for unneeded high doses of vitamin B_{12}. Vitamin B_{12} injections may be used as a placebo and are sometimes associated with weight-loss schemes.	

| Biotin | Necessary to four biotin-dependent enzymes: pyruvate carboxylase, methyl-crotonyl-coenzyme A [CoA] carboxylase, propionyl-CoA carboxylase, and acetyl-CoA carboxylase. Deficiency may be the result of altered intake, interference with metabolism, or increased demand. A deficiency may involve a genetic replacement in DNA and may be linked to altered fatty acid synthesis. Deficiency is not considered common. Def may be associated with renal hemodialysis; eating raw egg whites (which bind biotin and prevent absorption), treatment with anticonvulsants such as phenytoin, pyrimidine, and carbamazepine. Anticonvulsants may accelerate biotin breakdown. | Def: In the first three to five weeks, deficient intake results in dry skin; fungal infections, rashes, and fine, brittle hair or hair loss. Approximately one to two weeks later, neurologic symptoms develop. These include changes in mental status; mild depression, which may progress to profound lassitude; and somnolence. Possible neurological symptoms of adults include, depression, lethargy, hallucinations, memory failure, disorientation, confusion, psychotic episodes, and paresthesia of the extremities. Nausea, vomiting, and anorexia may also be present. "Malignant hiccups" (long-term, intractable hiccups as often as one per second) include depression, assaultive behavior, and suicidal ideation. Megadoses of biotin can result in symptom relief. Adults responsive to doses of 10 mg/day for three months. | Doses of free (unbound) biotin in the range of the estimated typical dietary intake (50 to 150µg/day) are shown to prevent the symptoms of biotinidase deficiency. | Food sources: meat, fish, poultry, egg, dairy, and some vegetables. |

(Continued)

Table 8 (Continued) Nutrients, the Brain, CNS, and Mental Status
See Appendix for References for Table 1.

NUTRIENT	BRAIN /CNS PHYSICAL STATUS	S/S MENTAL STATUS	DRI / DOSES	FOOD FACTORS
C – Ascorbic Acid	Vitamin C is a cofactor in conversion of dopamine to norepinephrine. Vitamin C may change receptor sensitivity, which interferes with stimulation of dopamine-sensitive adenylate-cyclase.	Deficiency results in personality changes, hypochondriasis, social introversion, anxiety, insomnia, potential suicidal ideation. Psychological changes occur earlier in depletion than physical changes. Personality changes occurred at 1.21–1.17 mg/100 ml whole blood ascorbic acid. Physical impairment occurred at 0.67–0.14 mg/100 ml. Changes in scores on MMPI: hypochondriasis, hysteria, depression, and social introversion. Reported symptoms: fatigue and weakness. Decreased excretion of vitamin C in 1–2 g load test in patients with schizophrenia.	RDA = 75–90 mg/day Replete body pool: 20 mg/kg or 500 mg vitamin C/day Smoking increases the need for vitamin C.	Food sources: highest fruits, cantaloupe, citrus fruits and juices, kiwi, mango, papaya pineapple, strawberries, raspberries, blueberries, cranberries, and watermelon. Highest vegetables: broccoli, brussels sprouts, cauliflower, green and red peppers, spinach, cabbage, turnip greens, other leafy greens, sweet and white potatoes, tomatoes and tomato juice, and winter squash.

D – Ergosterol	Possible function in regulating serotonin	Vitamin D level ≤50 nmol/L had higher likelihood of depression than those with serum vitamin D level was ≥75 nmol./L. Around 65% of patients with schizophrenia had vitamin D deficiency (Meta-analysis by Valipour)	RDA recommended Institute of Medicine (IOM) is 600 IU/day through age 70 and 800 IU/day for older ages. March 2015- Robert Heaney, M.D: "We call for the NAS-IOM to designate, as the RDA, a value of approximately 7,000 IU/day from all sources".	Food sources: fortified milk and other enriched foods. Skin exposure to Sun.
E – Tocopherol	A potent antioxidant in cell membranes also has anti-inflammatory functions. Vitamin C, present in the circulating plasma, restores vitamin E when the body needs antioxidant capacity recharged.	Vitamin E levels were inversely related to depression levels, even with an apparently adequate intake or when exceeding the recommended intake.	DRI for vitamin E is 15 mg/day. Tolerable UL is 1,000 mg/day. Extremely low fat diets may risk low vitamin E intake.	Food sources: vegetable oils, margarine, nuts (especially almonds), and seeds (especially sunflower seeds). Moderate sources include whole grains, egg yolk, collard greens, avocados, apples, and melons.

(Continued)

Table 8 (Continued) Nutrients, the Brain, CNS, and Mental Status
See Appendix for References for Table 1.

NUTRIENT	BRAIN /CNS PHYSICAL STATUS	S/S MENTAL STATUS	DRI / DOSES	FOOD FACTORS
Other				
Alcohol	Ethyl alcohol enters the brain within minutes of consumption and acts as a depressant affecting all nerve cells. The liver can detoxify ½ oz. of ethanol per hour. Alcohol interferes with the absorption of vitamins, as well as their function, especially B-1, B-3, B6, and folate.	Increased possibility of fetal alcohol effects (FAE) or fetal alcohol syndrome (FAS) in newborns. In addition to physical malformations, child may develop varying degrees of mental slowness or retardation, hyperactivity, mood disorders, and aggressive behavior. Alcohol intake is positively correlated with eating disorders and depression.	Alcoholic beverages containing added caffeine presents unusual risks to health and safety. In November 2010, the FTC (Federal Trade Commission) warned producers: "Consumers may not realize how much alcohol they have consumed because caffeine can mask the sense of intoxication". Several brands of these beverages, with 23.5 oz cans, provide alcohol content equal to three to seven light beers, as well as caffeine, taurine, guarana, and ginseng. http://www.ftc.gov/opa/2010/11/alcohol.shtm)	Sources: beer, wine, distilled liquor, some cough syrups. A "drink" is the amount of a beverage containing ½ oz. of pure ethanol.

| Caffeine | Caffeine belongs to the methylxanthine chemical family along with theophylline theobromine. Comparative potency of alkaloids: Highest: Theophylline Second Caffeine Third Theobromine (in chocolate) Estimates of plasma half-life of caffeine ranges between three to seven and a half hours. | In addition to physical effects, it improves concentration and extends intellectual efficiency. Single doses of caffeine up to 200 mg do not give rise to safety concerns. Habitual caffeine consumption up to 400 mg per day does not give rise to safety concerns for nonpregnant adults. | Coffee Tea Colas Energy drinks Chocolate (1–15 mg/oz in semisweet and milk chocolate) |
| Sugar | Humans produce opioids (the chemically active ingredient in heroin, cocaine and other narcotics) as a derivative of the digestion of excess sugars and fats. Endogenous opioids converge at the pleasure centers of the brain, the D2 dopamine receptors. | Out of control food consumption related to pain reduction centers focuses on the serotonin mechanisms in the brain. | |

(Continued)

Table 8 (Continued) Nutrients, the Brain, CNS, and Mental Status

See Appendix for References for Table 1.

NUTRIENT	BRAIN /CNS PHYSICAL STATUS	S/S MENTAL STATUS	DRI / DOSES	FOOD FACTORS
	Intense sweetness – not just refined sugar but also artificial sweeteners – surpasses cocaine as a reward in laboratory animals.			
Brain plasticity	Two types of brain plasticity: Functional plasticity: Refers to the brain's ability to move functions from a damaged area of the brain to other undamaged areas. Structural plasticity: Refers to the brain's ability to actually change its physical structure as a result of learning.	Plasticity is ongoing throughout life and involves brain cells other than neurons, including glial and vascular cells. Plasticity can occur as a result of learning, experience, and memory formation, or as a result of damage to the brain.		

See also Appendix for references of interest for Table 8: Nutrients, the Brain, CNS and Mental Status
See also Appendix for references for Chapter I – Why Integrate Nutrition Care and Mental Health Care?

Nutrition Detection Stories

But Officer, "I Would Never…"

"No, I would *never* hit her! I love her. I don't see why you're even asking me this", Mac said, as he tightened his grip on the arms of his chair. The officer assigned to Domestic Disturbances Division wondered if his tearfulness might be due to depression in this large, well-dressed man.

In a room down the Human Resources hallway came a similar reply, "Mac wouldn't hit me. He *loves* me. I don't even know why we're here". Biting her lip, Candace fidgeted, wishing for a cigarette.

Remembering the anonymous call from the Saint Veronica* prayer group, the employee resources social worker explained, "When you came to your appointment, and I saw all those bruises, I was concerned", looking at the large black and blue areas on the young woman's arms and neck. She continued, "Candace, I'm signing a referral for a physical examination by one of the in-house physicians. Would you prefer a female?"

"Open wide", Dr. Clark instructed, slipping the tongue depressor between her patient's lips. "When's the last time you saw the dentist?" she asked, as she checked off:

Gums: $\sqrt{}$ red/purple
$\sqrt{}$ bleeding
$\sqrt{}$ swollen/puffy

"Six months ago, and I didn't have any cavities", Candace replied once she could close her mouth.

Dr. Clark's attention went to Candace's arm, looking closely while stroking the fine hairs. *Every hair follicle is red at the base, she thought.* Does this hurt?" Dr. Clark asked, pressing her upper arm, looking for any signs of wincing. "How about this?" moving Candace's bent knee from right to left. Candace nodded reluctantly at the silent stab in her hip joint.

The doctor picked up her limp hand, noticing that beneath most nails were many tell-tale fine black lines from tip to the center of each fingernail. "See those dark lines under your fingernails?" Dr. Clark asked. "Those lines are from small capillary cell walls that are weak and bleeding".

The physician laid her stethoscope on the desk, picking up a clipboard and a pen. "Tell me about what you and your husband eat".

"Well, neither one of us really likes to cook, and Mac has lots of food allergies. We have shifts that don't start and end at the same time, so we don't get to eat meals together very often. I usually grab something out of the vending machines at the courthouse".

"Do you take any vitamins?"

"Not really. I read they just give you expensive urine".

"I'll need you to go down the hall for some lab work. Second door on the left. We'll both be notified of the results in five to seven days".

~~~

The official court report concluded, "Subject Candace Barr not found to be a victim of physical violence. Laboratory test confirms signs, symptoms, and diagnosis of advanced scurvy. Subject instructed to take a supplement of 250 mg/day of ascorbic acid daily to compensate for poor dietary intake and increased requirement for vitamin C due to smoking habit".

*Note to Court: Recommend adding orange juice and fresh oranges to court vending machines.

A. Z. Clark, MD

Ref: Sandstead, H. H. How to Diagnose Nutritional Deficiencies, Nutrition Teaching Aid No. 5. *Nutrition Today*, Washington, DC, 1969.

Choiceville, USA

Choiceville looked just as they remembered it as Dr. and Mrs. Michael Hamilton rolled into town in April, both ready to start the retirement phase of life: bike lanes, light traffic, lots of trees.

"Let's stop and get a copy of this week's C-Ville newspaper and come back later – all the outside seating is taken", suggested Helen as they approached the Bake and Brew, which appeared to be the local gathering spot.

"'MYSTERIOUS MALADY IN TOWN' REPORTS NEW WALK-IN CLINIC" shouted the headline, taking more than its share of the front page.

"Don't forget, I'm retiring," said Michael when Helen read it aloud.

Along with unpacking, and checking out the tech support center, the library and the mom 'n' pop grocery they made it back to the Bake and Brew the following week with their new neighbors Jack and Sue.

"It sure smells good in here; a good bakery is hard to beat".

"Yes, and they get all their flour from the local growers who have it milled just outside of town. It's all natural", explained Sue.

Lunch conversation included more details about life in Choiceville: "There is a great gym over on Second Street; the pickle-ball gang plays on Monday, Wednesday, and Friday from ten to noon; and the book group meets every other Tuesday evening at the community center. Oh, and the Farmer's Market is all day Friday, eight to four".

Michael and Helen exchanged smiles; they had made a good choice coming here. It was only 50 miles to a large mall, a hospital, and the symphony hall.

Two weeks later, Mike, missing the different kind of busy that medical practice brings, stopped by the Choiceville Walk-In Clinic and Pharmacy and introduced himself. After a tour by clinic physician Dr. Pete, Michael found himself offering, "If there is any way I can help you out, let me know".

"Sure thing," Pete replied. *I wonder if he really means that?*

Helen gave him the usual daily rundown over dinner. "I've never seen so many people on some kind of diet – the men as well as the women. I don't see how they do it".

"What do you mean?" asked Michael to keep up his end of the conversation.

"Almost everyone orders the half-sandwich special, washes it down with a local beer, and finishes up with coffee and one of Mrs. Sunderson's fantastic goodies. Her lemon tarts are the absolute best – not that I get one every day. Some days they run out by one o'clock".

"Do they exercise?" It seemed like the logical thing to ask to follow his wife's train of thought.

"You wouldn't believe how many miles of biking and jogging some of those women log every week! They say their husbands get even more! Although they say the slope of the sidewalks is difficult to manage some days".

The headline of the *C-Ville Herald* that week read "DR. PETE BAFFLED BY LOCALS' COMPLAINTS".

Michael dropped by the newspaper office before keeping his appointment at the clinic.

"But this is big local news", said the only reporter in the office when asked if he knew about HIPPA (Health Insurance Portability and Accountability Act). I don't name names. Everybody wants to know what's going on. "

As a matter of fact, so do I, thought Michael as he walked the block to the clinic.

"As a matter of fact, so do I. We've been inundated with new patients. I'm looking at every possible cause I can think of," exclaimed Pete shortly after Michael arrived. "You're years of experience ahead of me. What does this cluster of symptoms sound like to you?"

Dr. Pete described the common complaints to his colleague. "At least ten percent of the population have come in saying they had unusual muscle weakness in the thigh muscle when trying to go uphill on their bicycle or even with walking; they notice a shimmering sensation in their vision and/or loss of color vision, and most have numbness, prickling, and either burning or coldness in their feet. Even more people complain of fatigue, saying they are "so tired all the time" and many have diarrhea".

Dr. Hamilton sorted out symptoms and his knowledge of epidemiology. "Is alcohol consumption a problem in Choiceville?"

"Well, the Bake 'n' Brew is pretty busy, and the brewery a mile or so south of town has a 24-hour tasting room with no charge".

Is the bakery part of the business busy too?

"You bet – breakfast muffins, pastry, cookies; you name it. They even have customers from out of town".

"My wife says a lot of the people she meets are on a diet. What kind of diets do you hear about?

"A variety. Mostly low fat. Everyone is trying to cut calories and exercise to lose weight".

A 20-plus-year-old memory stirred to life in Dr. Michael – something about the Cuban Embargo in the 1990s.

"Didn't I hear something about the flour mill that supplies the bakery?"

"Right," explained Pete. "People like the fact it is all natural, straight from the mill to the bakery. People who live in Choiceville really go for the healthy foods. Most of them want to get all their nutrition from foods, lots of vegetables, not from a vitamin pill. The Farmers' Market does a booming business. A lot of patients are mostly vegetarian. That bakery is many people's one weakness".

"Let me think about this, and we can talk again tomorrow". Dr. Hamilton had a few things to check out.

Thank goodness for the Internet, Michael thought as he searched for information on the 1993–1994 international report about the consequences of the Cuban Embargo and for the state laws on fortification of flour. *Aha!*

He took his *Assessment of Nutritional Status* book with him to the clinic, arriving before the doors opened for patients.

"Here's what I found".

Pete was all ears.

"As far as I can tell, all the symptoms you're seeing can be explained by a deficiency of the water-soluble vitamin thiamine – also known as vitamin B-1. So, many factors are converging for the residents of Choiceville: alcohol and sugar intake increase the need for thiamine, intake is lower here because the usual fortification of flour isn't required and isn't done in the local mill, most fruits and vegetables aren't high in thiamine, and people here are not inclined to take supplements – so it all adds up. Thiamin is necessary for energy production and important for the nervous system, so you can see how this contributes to the signs and symptoms you are seeing in the clinic".

"Something very similar happened to 50,000 citizens of Cuba when the international embargo drastically reduced the country's food supply. In Choiceville, citizens did it with their own choices".

The next day's headline read, "VITAMIN B-1 NOW IN STOCK IN THE C-VILLE CLINIC AND PHARMACY. (For more information see the new Choiceville Clinic and Pharmacy website or contact the new webmaster, Dr. Michael Hamilton)".

###

There's a Method to Her Madness

Louise smiled as she remembered telling Dr. Jackson during her job interview, "Sure, I can get things organized around here". She spit out her bubble gum and squeezed her hair back into the rubber band from the newspaper. *I wonder if he'll notice my new system today?* She settled in her chair, ready to greet the stream of patients arriving to see their psychiatrist.

Agnes was the first patient in the door that morning, her usual agitation plainly evident. *Got to get her settled in an examination room to keep the reception room calm. She'll get her second injection in the series today.*

Louise lined up the medical records to match the patients as they arrived.

Arthur labored his way across the carpet to sign in. *Last time he opted for pills instead of an injection.*

Next in line, Alexis shrugged and replied, "I don't really care", when asked if Arthur could go ahead of her, even though her appointment time was first.

Carl entered without looking around, signed his name without saying a word, and sat down. He leafed through the *Daily Telegraph*, glanced at the *National Geographic*, flipped through *People* magazine, and then started pacing.

Glancing at Eileen, Louise put the box of tissues on the counter in plain sight, *I wonder how long she can sit today without bursting into tears.*

They were soon joined by Iris, who barked, "You have the very same magazines you had the last time I was here". She stomped to a seat and plopped down.

Then Mimi arrived. *She might not even remember Dr. Jackson's name, much less whether the magazines were the same or not.*

When Max walked in the door, Louise wondered, *Will he be "up" or "down" today?* Max approached the desk, signed in, and took a seat, not looking Louise in the eye.

The reception room filled with energy when Phil came bouncing in with a big smile and flourished a small bouquet of flowers, which he gave to Louise. *What a change from the last time he was here and called me a moron.*

Louise snapped to attention when Pauline arrived. Leaning over the desk, Pauline whispered, "What's that big shaggy animal standing behind you?" Louise led her directly to the examination room they kept open for instances like this. "Let me get you a cup of water to sip while you wait", she said. She looked at her watch. *It takes her daughter about five minutes to park the car and get up to "Psychiatry" on the second floor.*

She indicated the corner chair to Sam after he checked in, knowing he would probably fall asleep while he waited. *Had he been "awake at all hours of the night" again?*

"Is he even here?" asked Suzanne, motioning toward the picture of Dr. Jackson hanging in the reception area. Louise just nodded. Raising her eyebrows, Suzanne pursued the issue, "Would you tell me if he wasn't?"

~ ~ ~

Running a little late after a full morning, Dr. Jackson came into the reception area. Leaning on the counter he asked, "How on earth did you manage to get all my vitamin B-12 patients scheduled for the same morning?" The labels on the files weren't in view as Louise finished refiling the morning medical records into her new system:

B-12 Deficiency

Agnes – agitation
Alexis – apathy
Arthur – anemia
Carl – concentration

Eileen – emotional instability
Iris – irritability
Mimi – memory problems
Max – mood swings
Phil – personality changes
Pauline – psychosis
Sam – somnolence
Suzanne – suspiciousness

The Camp Conundrum

As he sat waiting in the expansive living room, the thoughts of the director of Camp Opportunity, Chuck Hopewell, swirled through his head: *How could this happen? To me? What will they say? I hope there won't be any lawsuits. It's not my fault!*

As they arrived in pairs or singly, the parents discussed and rediscussed the weather, silently asking themselves *How could such a thing happen at a camp? And a high-priced, all-summer camp, at that.* Chuck mustered high-fives to parents he knew better than others.

Dr. Johnson, the only functional medicine physician in town, arrived – casual in his jeans and Nike's, carrying his briefcase, and wearing a sober expression. During group introductions, he mentally matched up his recent patients with the right set of parents. *Heather Carruthers, who didn't respond to the antidepressants; Todd Scott, the tennis player with a dark scaly skin condition; and the twins Isaac and Izzie, whose mother kept saying, "Not MY twins! They always pay attention, and they don't pick fights!"* He thought about conversations with colleagues as they tried to figure out what they'd labeled "the Camp Conundrum".

Chuck silently rehearsed the answers to questions he had answered earlier, in case they came up again: *Yes, sunscreen is always available – at the boat dock, at the snack bar, and at the start of the hiking trail. Yes, hot food is always kept over 165 degrees F and cold foods are refrigerated or kept on ice. Yes, we keep an eye and ear out for bullying, and a counselor is always on duty.* He had been warned about, but was not prepared for, demanding parents when he started getting the telephone calls in August. Mrs. Carruthers had been the most upset, "Heather just couldn't quit crying the last few times she called. I couldn't tell exactly what the problem was. Are there some personality things going on up there? Some teens can be so mean to each other".

Chuck thought back over the 12 weeks, which had flown by. The kids were getting good suntans. There were few late escapades or arguments about bedtimes – the campers were obviously tired. The only problems reported by camp staff were that the supply clerk reported they were running out of toilet paper earlier than past summers, and the cabin counselors commented that kids seemed more nervous and anxious than usual about school starting.

Dr. Johnson stood up, clearing his throat, "Welcome, everyone. I want to start by passing out samples of the camp menus [see handout below].

"As you know, campers can eat as much or as little as they want, and some teens are very selective; some follow a vegetarian diet. Some come to camp wanting to lose weight, and there is a weekly weigh-in through the summer. As Chuck can tell you, Camp Opportunity aims to provide popular, attractive, safe meals. Beverages and desserts are purely optional choices. Menus are planned by a nutritionist with a diploma from the Eat Well School of Cuisine.

"What we're seeing in your children, after physical exams, interviews, and several panels of laboratory tests, is a final diagnosis of pellagra – a deficiency of the vitamin niacin. It has a variety of symptoms and is the result of an inadequate vitamin intake over the 12 weeks of summer camp".

Parents looked at each other with raised eyebrows, and several reached for their cell phones to access the internet. Mr. Scott, a PE teacher, nudged his wife. "No wonder Todd kept saying he was so tired after a match; if I remember correctly, niacin is required for energy production in all muscle cells".

Dr. Johnson continued, "The classic symptoms of pellagra are the three Ds: diarrhea, dementia, and dermatitis. The dermatitis is usually seen as a darkening, or redness, on areas exposed to the sun. Personality and mood changes, like depression, are among the earliest symptoms of borderline and acute deficiencies. There is generally no decline in intellectual or learning ability".

"An intake of 8.8 mg of niacin equivalents in a 2,000-calorie-per-day diet is associated with classic symptoms of niacin deficiency. The camp menus you have in front of you provide:

Total for Day 1: 6.6 mg niacin + 2.8 niacin equivalents from tryptophan.
Total for Day 2: 8.3 mg niacin + 4.16 niacin equivalents from tryptophan".

"Each camper has been prescribed a supplemental dose of niacin by his doctor, to be taken for two months. Retesting is recommended at that time to make sure optimum replenishment has been reached. I suggest that next year the camp menus be analyzed by computer to check the nutrient content. Any questions?"

Several hands shot up. Several parents spoke at once. "Could this have been prevented?"

"Yes, my very question".

"That's what I want to know".

Dr. Johnson explained, "Of course it depends on what each individual eats. The right choices have to be available, but ultimately, each camper chooses what he actually consumes". The thought ran through his mind, *There was no way of knowing if there was any genetic predisposition in this crowd; better just leave that alone.*

More questions came from the parent group: "How long do the symptoms last?"

"Do they go away entirely?"

"If each child takes the supplement recommended along with adequate meals, everyone should be symptom-free in two to three months". Dr. Johnson felt the tension in the room go down a notch or two.

Irate questions flew at Chuck, "How could you let this happen?"

"What are you going to DO about this?"

"Yeah, Chuck?"

"Chuck, is next year going to be any better?"

Shifting in his chair, Chuck drew a deep breath. *Did his nutritionist do an analysis? How much would a computer program cost? They didn't cover this in Camp Management 201!*

His mind churned out options, *How much more will it cost to add more fish, chicken, and beef to next year's menus ? I could add milk and yogurt to the evening snack cart, that would add niacin. Maybe I should add a vitamin to the daily charge for every camper.*

Rising, Chuck looked each parent in the eye, "My wholehearted apology to you and your children for not being aware of this as a potential problem. You can bet I'll be planning some steps to prevent anything like this from ever happening again. As I get new plans in place, I'll be sending out a letter to each and every family to assure you that things will be definitely be different next summer. Thank you for your involvement in solving this unfortunate situation".

He closed his briefcase, the folder with next year's camp applications inside. *Better just wait to hand these out.*

Camp Opportunity – Sample Menus

Summer 2016

Day 1

Breakfast	Lunch	Dinner
Fresh orange Poached egg Oatmeal bread toast with margarine, jelly Coffee, sugar, creamer	Lettuce, green leaf Broccoli, raw Garbanzo beans Cheddar cheese, grated Italian dressing Wheat crackers Gelatin with fruit cocktail Chocolate cake Iced tea, honey	Shrimp, breaded, fried Baked potato Cabbage, raw, shredded Mayonnaise (soy oil, salt) Strawberries, raw, halves Soda

Day 2

Breakfast	Lunch	Dinner
Cantaloupe Quaker natural granola with raisins Coffee, sugar, creamer	Soup: vegetable or clam chowder Oat bran muffin, margarine Celery stuffed with cream cheese Banana Iced tea, honey	Taco shells, baked, 2 Ground beef, 2 oz Lettuce, iceberg Tomato, raw, chopped Cheddar cheese, grated, 1 oz Refried beans Apple, raw Lemonade, from concentrate Ice cream, vanilla

---

Total for Day 1: 6.627 mg niacin + 2.8 niacin equivalents from tryptophan.
Total for Day 2: 8.311 mg niacin + 4.16 niacin equivalents from tryptophan.

###

Step Right Up – Play "Life Today"

The contestants were trying not to do two things: not to glance at their competitors and not sweat as they thought about the big prize. Todd could see that $250,000 grand prize wiping out his college debt, and then some. Elaine was crossing her fingers that tomorrow she wouldn't have to worry about the bill for that new stent in her carotid artery. *Costa Rica, here I come*, said Karen to herself. Bill wondered, *Exactly how much is the balance on Mom and Dad's mortgage?*

Host of *Life Today*, Maggie Thomas's finger was poised on the keys that unlocked the next phase of the game: **Level Six: Health Through Nutrition**.

She reminded the contestants, "Remember, if you make two incorrect guesses, you are out of the game; in some cases, it's better to answer, "Don't know". She opened question number one for the Level Six nutrient.

"Are you ready for Question 1?" Four heads nodded. "Only 25% of Americans consume enough of this nutrient. Which nutrient is this?" Nearing the final second, four "Don't Know" answers lit up.

"Alright, let's go on to Question 2: Coffee, carbonated beverages and alcohol all deplete the body of this nutrient, although through different mechanisms. Can you name this nutrient?" Again, the maximum allowed pause went by with no guesses.

"Can anyone answer Question 3?: Muscle cramps, coronary spasms, and irregular heartbeat may occur if you are deficient in this nutrient. Which nutrient could it be?"

Remembering his cramps during the marathon last month, Todd's light went on: "Potassium," he called out.

"A good try. I'm sorry, but that's not the correct answer". Maggie prepared for the next question.

"Question 4: This mineral, a muscle relaxant, needs to be balanced with calcium intake, which makes muscles contract. The balance is hard to determine because a blood test often does not give an accurate picture of this mineral in the rest of the body". Karen made a wild guess based on an advertisement she had read online. "Zinc," she answered.

"Excellent try. I'm sorry, but zinc is not the correct answer," replied Maggie, pressing the key for question five.

Question 5 appeared on the screen: "Being linked to both plaque and stiffening of arteries, insufficient levels of this nutrient is a better predictor of heart disease than either cholesterol levels or high blood pressure". Four answer lights lit up, "Don't know".

"Next we have the final question in Level Six of Health Through Nutrition". Maggie announced, "Are you ready?" She could see the gleam of sweat on four foreheads. "Question 6: A good food source of this mineral is green leafy plants because this mineral is the central molecule in chlorophyll".

All four answer lights blazed and four voices shouted out in unison, "Magnesium!"

Maggie's voice boomed over the cheers of the studio audience, "We have *four* winners! An exceptional race! Congratulations to each and every one of you!"

Between hugs, OMGs, and tears, Karen wondered if she could be ready tomorrow for boarding a flight to Costa Rico; Bill pulled out his cell phone to call his Mom and Dad before they got in bed; Elaine breathed a prayer of thanks for being able to pay those bills after all; Todd mentally went down the list of things his son had put on his list for Santa Claus.

"Each contestant takes home $62,500 and has a chance to play *Life Today* again next week! If you would like to be a contestant, send your entry form to the address on your screen. Til next week, good night from all of us at *Life Today*".

Too Much or Not Enough?

Within two minutes of the "newbies" starting their ER rotation at Riverside General Hospital, nervously anticipating whatever the day might bring, the swinging door burst open. Two young men in shorts and Nikes came staggering in, with a third man sagging between them, the words "2017 Marathon" emblazoned on their T-shirts. Team One sprang into action strapping No. 45 onto a gurney and headed toward Bay 1.

Meanwhile, Team Two hurried to the door to meet a middle-aged woman and an elderly woman slumped in a wheel chair who had a nurse beside them.

"Right this way. What seems to be the problem?" asked the resident leading the way to Bay 2.

As the door swung closed, it immediately reversed. A young woman, with what appeared to be a parent on each side, approached the admissions desk. "Something is really wrong; we need to see a doctor," said the man holding the woman's right arm. The nurse supervisor led the group to Bay 3.

~ ~ ~

"Let's sum up the cases from today". Dr. Voda motioned the ER team members to sit down for the usual briefing after being relieved by the incoming shift. "Team One, fill us in".

Russ, the leader of Team One, spoke up, "Our patient was a 24-year-old male in reportedly good health. His friends told us that at one point during the marathon, he got confused and started running in the wrong direction. As they tried to get him turned around, he started holding his head, threw up, and then collapsed on the ground".

"Was he stopping at the water stations on the route?" asked Dr. Voda.

"Definitely. In fact, he apparently was quite thirsty according to his friends", added Amy, the nurse on Team One.

Moving on, Dr. Voda asked, "Team Two, tell us about the woman you treated".

The team leader, Lisa, reported. "She was an 86-year-old woman brought in sitting in a wheelchair by her daughter and the nurse who does home visits". Lisa repeated what the nurse and daughter told her. "Mrs. Louis was doing okay until a few weeks ago when she started having kidney problems – bladder leakage and needing frequent help to the bathroom. She didn't have much appetite – just ate fresh fruits and vegetables that were cooked well done so she could chew them. This morning she had cramps and then a seizure, so they thought they should bring her in to get checked".

Dr. Voda asked "Did the daughter say how she cooked or seasoned the fruits and vegetables that she fed her mother?"

"I really didn't ask her", Lisa replied.

"Let's hear about our third admission this morning," Dr. Voda turned to Team Three.

Team leader, Rob, spoke for the team from Bay C. "Her parents said she went to a party last night. They heard her come in and then heard her throwing up in the bathroom. When they went to check on her she was confused and disoriented, then fainted. The patient is on several medications, but they don't think she ever does any drugs at parties. They didn't want her to get dehydrated from vomiting, so they encouraged her to drink plenty of water".

Dr. Voda began the review, "What is the common factor in these three cases?"

All hands shot up, "Water!"

"Too much or too little water? What is the other factor?"

Team members were on their toes, some replying "Sodium", others replying "Salt".

"Let's make a list". Dr. Voda stood up and wrote on the white board: *Hyponatremia – Symptoms and Causes.* One by one, team members added to the list of symptoms and factors related to very low levels of sodium in the blood:

Hyponatremia

Symptoms

Nausea, vomiting, headaches

Muscle weakness, cramps

Confusion, disorientation

Seizures, coma

Decreased pulse rate, bradycardia

Causes

Excessive rehydration after exertion

Low intake and high excretion of sodium

Vegetables and fruits are low in sodium

Thirst induced by prescribed or illicit drugs:

SAIDS, MDMA (ecstasy), antipsychotics

Medical conditions that may alter

metabolism: kidney disease, congestive

heart failure, diabetes, liver disease

Dr. Voda continued to write: *"Factoids to Remember in Case Anyone Ever Asks"*:

- Healthy kidneys are able to excrete approximately 0.84 to 1.04 quarts per hour (800 milliliters to 1 liter) of fluid water.
- A normal sodium level is between 135 and 145 milliequivalents per liter (mEq/L) of sodium. Hyponatremia occurs when the sodium in your blood falls below 135 mEq/L.
- The IOM recommends that a healthy adult drink 78–100 ounces (about 9–13 cups) of fluids per day, on average. Water needs vary with age, sex, weather, activity level, and overall health; there is no exact formula.
- If fluid outside the cells has an extremely low amount of sodium and other electrolytes in comparison to fluid inside the cells, this causes fluid to move into the cells, causing the cells to swell. The cells in the brain may swell to the point where blood flow is interrupted, resulting in cerebral edema and central nervous system dysfunction. Interference with the central nervous system could result in seizures, brain damage, coma or death.

Everyone copied everything off the board; heads nodded and fingers kept scribbling as Rob commented, "I know I know, but I'd better write it down anyway, just in case".

Note: Voda = water in Croatian, Czech, Russian, and Slovak

The Nonanonymous Support Group

The sign on the door matched the paper he had in his hand "HIFO-SG meeting today 4:00–5:00 PM", so Chris knocked. A bald woman answered the door.

"Come in, I'm Becky. Welcome to the Hair Is Falling Out Support Group".

Hesitating as he looked around the large comfortably furnished room with overstuffed chairs, Chris was greeted by an outstretched hand, "Hi, I'm Louis, here bemoaning the genes I've apparently acquired from my dad. How about you?"

"I'm trying to figure out why I'm still losing hair even though my successful cancer therapy was finished several months ago", Chris confided.

"That's great to hear about the cancer but not so great about the hair. Let me introduce you around", said Louis, leading the way.

"Chris, this is Malao, our youngest member. How is the trial of a head of flowing hair instead of corn rows going, Malao?"

She answered, "It hasn't been long enough to tell yet. Hair only grows about six inches a year, and I changed my hairdo only about five months ago".

Turning to a middle-aged woman standing on their right, Louis said, "Rebecca let me introduce you to a visitor, Chris". Chris backed away as a reflex against the strong smell of garlic surrounding her.

"Nice to meet you", Chris said, turning away and bumping into a man of about 25 or 30 with long stringy hair, who introduced himself.

"Hi, I'm Nick. They tell me my hair problems are because I'm a vegan and don't eat enough protein, even though I eat a lot of nuts. I've been trying eggs and milk. At least the animals don't get killed. Here's my mom, Alicia. Mom, this is Chris".

Alicia introduced herself according to group custom – by being upfront with her issues about hair loss. "I'm not only losing my hair, but also my nails. I take vitamins, minerals, and I drink Ensure a couple of times a day… even doubled up on the minerals since I hear they are good for fingernails, but so far, they don't seem to be helping".

Chris wandered over to the refreshment table and helped himself to some tea and picked out a few Brazil nuts, his favorite.

An older, thin, fragile man with a noticeable tremor and who was nearly bald introduced himself, "Hello, I'm Reggie. I'm not only losing hair, but trying to figure out if it could be related in any way to the statins and warfarin medications I'm taking. Seems like everything affects everything else these days. After a pause, he continued, "Here, let me introduce you to Dale, who was new last month. Dale, this is Chris".

"Nice to meet you, Chris". Dale continued, "I've been wondering if hair falls out from too little or too much of something. I went to my

PCP, and he wanted a clipping from my big toenail to test! Have you ever *heard* of such a thing?" Chris hadn't.

They all turned to listen to a woman on the far side of the room standing and waving some pieces of paper, speaking in a loud voice. "Attention! Attention! I found a great resource for everybody! It tells a lot about how hair is affected by having too much selenium! For one thing, most of us need about 55 micrograms/day and one little Brazil nut has almost twice that."

Chris nearly choked, remembering the nut dish beside his recliner.

The woman continued, "If you eat five Brazil nuts in a day, you're over the upper safe limit: four hundred micrograms! Or if you combine too many supplements that all have selenium, it could be toxic".

Alicia's thoughts raced, "I wonder exactly what those labels add up to".

The loud woman continued, "Guess how you can tell? You start to smell like garlic or have a metallic taste in your mouth! *And* your hair starts falling out!"

Chris wondered, "Hope Rebecca heard that".

The woman continued, "And it says right here that too much selenium can affect how some medications act in your body !"

Reggie turned his hearing aid up and moved closer to the woman speaking.

"Other toxicities, like too much vitamin A, can make you lose your hair, or even nutritional deficiencies, like iron". Waving the papers, she continued, "I brought a bunch of copies, if anyone wants one, or you can look it up on the website for the Office of Dietary Supplements at https://ods.od.nih.gov/factsheets/Selenium-Consumer/#h8 or look up selenium at Consumer Labs at https://www.consumerlab.com/RDAs/".

She continued, "This is a lot to pay attention to, folks! These are things we all need to know". Seeing heads nodding across the room, she waved the handouts once more and finished with, "Shall we start a new discussion group on this?"

###

Appendix

The following Forms and Assessments are accessible on the Routledge
website: https://www.routledge.com/9781032842721

List of Tables

List of Forms and Assessments

Dietary Guidelines for Americans 2020–2025 Executive Summary (Excerpt) The Guidelines: "Make Every Bite Count with the Dietary Guidelines for Americans."

Here's how:

1. **At every life stage – infancy, toddlerhood, childhood, adolescence, adulthood, pregnancy, lactation, and older adulthood—it is never too early or too late to eat healthfully**.
   - For about the first 6 months of life, exclusively feed infants human milk. Continue to feed infants human milk through at least the first year of life, and longer if desired. Feed infants iron-fortified infant formula during the first year of

life when human milk is unavailable. Provide infants with supplemental vitamin D beginning soon after birth.

- At about 6 months, introduce infants to nutrient-dense complementary foods. Introduce infants to potentially allergenic foods along with other complementary foods. Encourage infants and toddlers to consume a variety of foods from all food groups. Include foods rich in iron and zinc, particularly for infants fed human milk.
- From 12 months through older adulthood, follow a healthy dietary pattern across the lifespan to meet nutrient needs, help achieve a healthy body weight, and reduce the risk of chronic disease.

2. **Follow a healthy dietary pattern at every life stage.** Customize and enjoy nutrient-dense food and beverage choices to reflect personal preferences, cultural traditions, and budgetary considerations. A healthy dietary pattern can benefit all individuals regardless of age, race, ethnicity, or current health status. The Dietary Guidelines provides a framework intended to be customized to individual needs and preferences, as well as the food ways of the diverse cultures in the United States.

3. **Focus on meeting food group needs with nutrient-dense foods and beverages and stay within calorie limits.** An underlying premise of the Dietary Guidelines is that nutritional needs should be met primarily from foods and beverages – specifically, nutrient-dense foods and beverages. Nutrient-dense foods provide vitamins, minerals, and other health-promoting components and have no or little added sugars, saturated fat, and sodium. A healthy dietary pattern consists of nutrient-dense forms of foods and beverages across all food groups, in recommended amounts, and within calorie limits. The core elements that make up a healthy dietary pattern include:

- Vegetables of all types – dark green; red and orange; beans, peas, and lentils; starchy; and other vegetables
- Fruits, especially whole fruit
- Grains, at least half of which are whole grain
- Dairy, including fat-free or low-fat milk, yogurt, and cheese, and/or lactose-free versions, and fortified soy beverages and yogurt as alternatives

- Protein foods, including lean meats, poultry, and eggs; seafood; beans, peas, and lentils; and nuts, seeds, and soy products
- Oils, including vegetable oils and oils in food, such as seafood and nuts

4. **Limit foods and beverages higher in added sugars, saturated fat, and sodium, and limit alcoholic beverages.** At every life stage, meeting food group recommendations – even with nutrient-dense choices – requires most of a person's daily calorie needs and sodium limits. A healthy dietary pattern doesn't have much room for extra added sugars, saturated fat, or sodium – or for alcoholic beverages. A small amount of added sugars, saturated fat, or sodium can be added to nutrient-dense foods and beverages to help meet food group recommendations, but foods and beverages high in these components should be limited.

Limits are:

- Added sugars – Less than 10 percent of calories per day starting at age 2. Avoid foods and beverages with added sugars for those younger than age 2.
- Saturated fat – Less than 10 percent of calories per day starting at age 2.
- Sodium – Less than 2,300 milligrams per day – and even less for children younger than age 14.
- Alcoholic beverages – Adults of legal drinking age can choose not to drink, or to drink in moderation by limiting intake to 2 drinks or less in a day for men and 1 drink or less in a day for women, when alcohol is consumed. Drinking less is better for health than drinking more. There are some adults who should not drink alcohol, such as women who are pregnant.

https://www.dietaryguidelines.gov/sites/default/files/2020-12/
DGA_2020-2025_ExecutiveSummary_English.pdf

Common Abbreviations

AA	Amino Acid
AN	Anorexia Nervosa
AND	Academy of Nutrition and Dietetics
ANS	Assessment of Nutritional Status
B-1	Thiamin
B-2	Riboflavin
B-3	Niacin
B-6	Pyridoxine
B-9	Folic Acid, Folacin
B-12	Cobalamin
BHN	Behavioral Health Nutrition
BMI	Body Mass Index
BN	Bulimia Nervosa
Ca	Calcium
CHO	Carbohydrate
CNS	Certified Nutrition Specialist
Cr	Chromium
Cu	Copper
D-2	Ergocalciferol
D-3	Cholecalciferol
DIFM	Dietitians in Integrative and Functional Medicine
ED	Eating Disorder
EFA	Essential Fatty Acid
Etoh	Ethyl Alcohol, Distilled Alcoholic Beverage
HTN	Hypertension
Hg	Mercury
IBS	Irritable Bowel Syndrome
IOM	Institute of Medicine
K	Potassium
MAOI	Monoamine Oxidase Inhibitor
Mb	Molybdenum
Mg	Magnesium
MSUD	Maple Sugar Urine Disease
NFPE	Nutrition-Focused Physical Examination
Pb	Lead
Phe	Phenylalanine – an essential amino acid
PKU	Phenylketonuria

RDN Registered Dietitian Nutritionist
Se Selenium
SIBO Small Intestine Bacterial Overgrowth
Vitamin C Ascorbic Acid
Zn Zinc

Website Resources

Academy of Nutrition and Dietetics Referrals website https://www.eatright.org/find-an-expert
Alcoholics Anonymous www.aa.org
BMI Calculation Online
Calorie Calculator https://www.mayoclinic.org/healthy-lifestyle/weight-loss/in-depth/calorie-calculator/itt-20402304 Dietary Guidelines: for information and tool kit regarding 2015–2020, https://health.gov/dietaryguidelines/
Dietetic Practice Groups Websites
Behavioral Health Nutrition https://www.bhndpg.org/
Dietitians in Integrative and Functional Medicine https://integrativerd.org/ FeedingAmericahttps://www.feedingamerica.org/find-your-local-foodbank
Food Composition Tables https://www.nal.usda.gov/fnic/food-composition
Food Pantries, Soup Kitchens: Serving communities directly https://www.foodpantries.org/, https://www.cccstudentmentalhealth.org/resource/college-and-university-food-bank-alliance-cufba/
HHS – U.S. Department of Health and Human Services http://www.hhs.gov/
IAEDP – International Association of Eating Disorder Professionals http://www.iaedp.com/about-us/
Integrative Medicine for Mental Health http://www.immh.org/
ISNPR – International Society for Nutritional and Psychiatric Research http://www.isnpr.org/
Linus Pauling Institute at Oregon State University Micronutrient Information Center https://lpi.oregonstate.edu/mic
National Alliance on Mental Illness 3803 N. Fairfax Dr. Suite 100, Arlington, Virginia 22203 https://www.nami.org/

NIAAA National Institute on Alcohol Abuse and Alcoholism. https://www.niaaa.nih.gov/, https://www.niaaa.nih.gov/publications/brochures-and-fact-sheets

NCIIH – National Center for Complementary and Integrative Health https://nccih.nih.gov/health/integrative-health

NIH – National Institute of Health https://www.nih.gov/

NIMH https://www.nimh.nih.gov/health/publications/index.shtml Health topics and educational resources http://www.nimh.nih.gov/index.shtml

Office of Dietary Supplements https://ods.od.nih.gov, https://ods.od.nih.gov/factsheets/list-all/, https://ods.od.nih.gov/factsheets/Folate-HealthProfessional/

Pub Med – U.S. National Library of Medicine National Institute of Health http://www.ncbi.nlm.nih.gov/pubmed

SAMHSA (Substance Abuse and Mental health Services Administration) https://www.samhsa.gov/about-us

SNAP: Supplemental Nutrition Assistance Program (food stamps) www.fns.usda.gov

Topics – http://www.samhsa.gov/topics

USDA – U. S. Department of Agriculture www.usda.gov
Dietary guidelines http://www.cnpp.usda.gov/dietaryguidelines
Evidence library http://www.cnpp.usda.gov/nutritionevidence library
National Agricultural Library http://fnic.nal.usda.gov/food-composition
Food Composition Tables http://fnic.nal.usda.gov/food-composition/vitamins-and-minerals, http://fnic.nal.usda.gov/food-composition/macronutrients

USDA Dietary Reference Intakes – https://www.nal.usda.gov/fnic/dietary-reference-intakes

References for Chapter I – Why Integrate Nutrition Care and Mental Health Care?

1. Garland, Malcolm R., Brian Hallahan, Mairead McNamara, Philip A. Carney, Helen Grimes, Joseph R. Hibbeln, Andrew Harkin and Ronan M. Conroy. Lipids and essential fatty acids in patients presenting with self- harm. *British J. Psychiatry*. 2007; 190:112–117.

2. Keys, Ancel et al. *The Biology of Human Starvation*. 1950. University of Minnesota Press, Minneapolis.

3. Kalm, Leah M. and Richard D Semba. They starved so that others be better fed: Remembering Ancel keys and the Minnesota Experiment. *J Nutrition*. 2005; 135:1347–1352.

4. Roy-Byrne, P., D.A. Gorelick and S. R. Marder. Unusual dietary habits in a patient with schizotypal personality disorder: Interaction of nutritional status and psycho-pathology. *J Psychiatric Treatment and Evaluation*. 1983; 5:67–69.

5. Sarris, Jerome, Alan C. Logan, Tasnime N Akbaraly, et al. Nutritional medicine as mainstream in psychiatry. *Lancet Psychiatry*. 2015, March; 2(3): 271–274. doi: 10.1016/S2215-0366(14)00051-0

6. U.S. Department of Health and Human Services and U.S. Department of Agriculture. *2015 – 2020 Dietary Guidelines for Americans*. 8th edition. December 2015. Available at http://health.gov/dietaryguidelines/2015/guidelines/

7. Office of Disease Prevention and Health Promotion. Scientific Report of the 2015 Dietary Guidelines Advisory Committee, Part A. Executive Summary. Page 2. odphpinfo@hhs.gov

8. Kinsman, Robert A. and James Hood. "Some behavioral effects of ascorbic acid deficiency." *Am J Clin Nutr*. 1971; 24:455–464.

9. Louwman, M. W. J., M. van Dusseldorp, J. R. Fons, et al. Signs of Impaired function in adolescents with marginal cobalamin status. *Amer J Clin Nutr*. 2000; 72(3):762–769.

10. Sliter, Kristine J. *Dietary Assessment in Standard Practice of Mental Health Professionals, Dissertation – College of Behavioral Sciences*. August 2014. Argosy University, Phoenix, AZ.

11. Hargrove, Emily J., Darlene E. Berryman, Jennifer M. Yoder, and Elizabeth. Assessment of nutrition knowledge and attitudes in preclinical osteopathic medical students. *J Amer Osteopathic Assoc*. 2017 October; 117:622–633. doi: 10.7556/jaoa.2017.119

Additional Reading

1. Campbell, T. Colin and Howard Jacobson. *Whole – Rethinking the Science of Nutrition*. 2013. Ben Bella Books, Dallas, TX.

2. Nagel, Thomas. The mind-body problem and psychoanalysis. *J Amer Psychoanalytic Assoc*. 2016; 64(2):389–403. doi: 10.1177/000306511 6647053

3. Quirk, S. E. et al. The association between diet quality, dietary patterns and depression in adults: A systematic review. *BMC Psychiatry*. 2013; 13:175. doi: 10.1186/1471-244X-13-175.

4. Siegel, Daniel. *Mind; A Journey to the Heart of Being Human*. 2017. W.W. Norton & Company, New York.

5. Kinsman, Robert A. and James Hood. "Some behavioral effects of ascorbic acid deficiency" *Am J Clin Nutr*. 1971;24:455–464.

6. Su, KP. Mind-body interface: The role of n-3 fatty acids in psychoneu-roimmunology, somatic presentation, and medical illness comorbidity of depression. *Asia Pac J Clin Nutr.* 2008;17(Suppl 1):151–467.
7. Hedges, Dawson W., Fu Lye Woon, and Scott P. Hoopes. Caffeine-induced psychosis. *CNS Spectr.* 2009;14(3):127–129.
8. Grubb, B. Hypervitaminosis a following use of high-dose fish oil supple-ments. *Chest.* 1990; 97:1260.
9. Hightower, J.M. and D. Moore. Mercury levels in high-end consumers of fish. *Environmental Health Perspectives.* 2003; 111(4): 604–608.
10. Tufan et al. Mood disorder with mixed, psychotic features due to vitamin B12 deficiency in an adolescent: case report. *Child Adol Psychiatry Ment Hlth.* 2012; 6:25. http://www.capmh.com/content/6/1/25
11. Parade Publications. "What Americans think about aging and health." *Parade Magazine.* New York, February 5, 2006:11.
12. http://www.nimh.nih.gov/statistics
13. Kertesz, S.G. Pellagra in 2 homeless men. *Mayo Clin Proc.* 2001; 76:315–318. doi: 10.4065/76.3.315

References for Table 2: Reports and Observations That May Be Related to Nutritional Risk

Kinsman, R.A. and J. Hood. Some behavioral effects of ascorbic acid deficiency. *Amer J Clin Nutr.* 1971; 24:455–464.

Oldham, Mark A. and Ana Ivkovic. Pellagrous encephalopathy present-ing as alcohol withdrawal delirium: A case series and literature review. *Addict Sci Clin Pract.* 2012; 7(1):12. Published online 2012 Jul 6. doi: 10.1186/1940-0640-7-12

Tu, H., Y. Wang, M. Niyyati, et al. Low red blood cell vitamin c concentra-tions induce red blood cell fragility: A link to diabetes via glucose, glu-cose transporters, and dehydroascorbic acid. *EBioMedicine.* 2015 Oct 3; 2(11):1735–1750. doi: 10.1016/j.ebiom.2015.09.049.

Valente, Maria J. and Neil Abramson. Easy Bruisability. *So Med J.* 2006; 99(4): 366–370.

References for Laboratory Assessment of Nutritional Status

Elmadfa, I and Meyse, A.L. Developing suitable methods of nutritional status assessment: A continuous challenge. *Adv Nutr.* 2014 Sep; 5(5):590S–598S. https://www.ncbi.nlm.nih.gov/pubmed/?term=ibrahim+elmadfa+-developing+suitable+methods+of+nutritional+status+assessments

Jones, A.D. Food insecurity and mental health status: A global analysis of 149 countries. *Am J Prev Med.* 2017; 52. http://doi.org/10.1016/j.amepre.2017.04.008

Koulouri, Olympia and Mark Gurnell. How to interpret thyroid tests. *Clin Med.* 2013; 13(3):282–286.

National Research Council (U.S.) Subcommittee. Chapter 6. Protein and amino acid. *Recommended Dietary Allowances*. Tenth Edition. National Academies Press. Washington (DC): National Academies Press (U.S.); 1989. http://www.advancedmolecularlabs.com/pre-workout-studies/the-optimal-timing-of-leucine-consumption/

Tabarki, Al-Hashem and Alfadhel. Biotin-thiamine-responsive basal ganglia disease. *Gene Rev.* 2013 https://www.ncbi.nlm.nih.gov/books/NBK 169615/11/21/2013

Walsh, W. J., L. B. Glab and M. L. Haakenson. Reduced violent behavior following biochemical therapy. *Physiol Behav.* 2004 Oct 15; 82(5):835–839.

Walsh, William J. *Nutrient Power: Heal Your Biochemistry and Heal Your Brain.* 2012. Skyhorse Publishing. New York.

Walsh, William J., et al. Elevated Blood Copper/ Zinc Ratios in Assaultive Young Males. *Physiol Behav.* 1997; 62(2):327–329.

Zaslavsky, Oleg, Shira Zelber-Sagi, James R. Hebert, et al. Biomarker-calibrated nutrient intake and healthy diet index associations with mortality risks among older and frail women from the Women's Health Initiative. *Am J Clin Nutr.* 2017. http://ajcn.nutrition.org/content/early/2017/04/18/ajcn.116.151530.abstract

References for the Nutrition-Focused Physical Examination

Lawrence, C.M. and Cox, N.H. *Physical Signs in Dermatology Color Atlas and Text.* Key Pharmaceutaicals, New Jersey. Mosby Wolfe, 1993.

McLaren, Donald S. A. *Colour Atlas and Text of Diet-Related Disorders.* 2nd Edition. Mosby-Year Book Europe Limited, Aylesbury, England. 1992, p. 89.

Mir, M. Afzal. *Atlas of Clinical Diagnosis.* W.B Saunders, Philadelphia, 1995.

Sandstead, H. H. How to diagnose nutritional deficiencies, Nutrition Teaching Aid No. 5, *Nutrition Today*, Washington DC, 1969.

Vitamin Manual; A SCOPE Monograph. The UpJohn Company, Kalamazoo, MI. 1965, p.15, Figure 4A (out of print) Walter Wilkins, MD, Jacksonville, Florida, and *Clinical Nutrition*, edited by Norman Joliffe, et al., Paul B. Hoeer, Inc., New York City, 1950.

Xu, Y.C., Vincent, J.I. Clinical measurement properties of malnutrition assessment tools for use with patients in hospitals: a systematic review. *Nutr J.* 2020; 19:106. https://doi.org/10.1186/s12937-020-00613-0

Selected References for the Gut, Microbiome, and Brain

- Relationship between the gut microbiome and brain function M. Hasan Mohajeri, Giorgio La Fata, Robert E. Steinert, and Peter Weber. *Nutrition Reviews* 2018. Vol. 76, No. 7: 481r–496r. https://doi.org/10.1093/nutrit/nuy009

- The microbiota–gut–brain axis in gastrointestinal disorders: stressed bugs, stressed brain or both? Giada De Palma, Stephen

M. Collins, Premysl Bercik and Elena F. Verdu. *Journal of Physiology-Topical Review* 2014. Vol. 592, No. 14: 2989–2997. https://www.health.harvard.edu/diseases-and-conditions/the-gut-brain-connection

- Diet-microbiome interactions in health are controlled by intestinal nitrogen source constraints. Holmes, Andrew, et al. *Cell Metabolism*. Jan 10, 2017. Vol 25, No. 1, 140–151. doi.org/10.1016/j.cmet.2016.10.021

- The gut microbiome and mental health: Implications for anxiety- and trauma-related disorders. Malan, Stefanie, Mireia Valles-Colomer, Jeroen Raes, Sian Hemmings. *Omics: a Journal of Integrative Biology*. 2017. Vol 21. No. 0. https://doi.org/10.1089/omi.2017.0077. https://www.researchgate.net/publication/318870864_The_Gut_Microbiome_and_Mental_Health_Implications_for_Anxiety-_and_Trauma-Related_Disorders

- Adaptation of the gut microbiota to modern dietary sugars and sweeteners. Sara C. Di Rienzi, Robert A Britton. *Advances in Nutrition* May 2020. Vol. 11, No 3: 616–629, https://doi.org/10.1093/advances/nmz118

Selected References for Table 8: Nutrients, the Brain, CNS, and Mental Status

Amino Acids

Prator, Bettina C. "Serotonin syndrome." *Journal of Neuroscience Nursing* 38(2) (2006):102–105.

Phenylalanine

Elsas II LJ and Acosta PB. *Clinical Laboratory Tests*, 1991; Springhouse Corp. Springhouse, PA. 486–487.

Elsas II LJ and Acosta PB. "Nutrition Support of Inherited Metabolic Disease". Ch 67. M Shils, JA Olson and M Shike. *Diet and Nutrition in Disease*. Eighth edition. Vo. 2. Lea and Febiger, Malvern, PA. 1994. p. 1155–1168.

Elsas II LJ and Acosta PB. *Dietitian Patient Education Manual, Vol I*. Aspen Reference Group, Aspen Publishers, Inc., Gaithersberg, Maryland, 1999: 11(1–11):4.

Rosenberg Leon E. "Inherited Disorders of Amino Acid Metabolism and Storage." Ch 352. *Harrison's Principles of Internal Medicine*, 13th edition, Isselbacher, et al, editors, McGraw-Hill, Inc., San Francisco, 1994. p. 2119–2122.

Rosenberg Leon E. *Illustrated Guide to /diagnostic Tests*. Springhouse Corporation, Springhouse, PA 1994: 218–220.

Sullivan Jill E. and PiNian Chang. "Review: Emotional and behavioral functioning in phenylketonuria." *J Ped Psychology* 24(3) *(*1999): 281–299.

Carbohydrates

American Medical Association, International Life Sciences Institute, and The Nutrition Foundation, Inc. "Diet and Behavior: A Multi-Disciplinary? Evaluation Symposium Panel Statement." *Nutrition Reviews* 42(5) (1984):200–201.

Comisarow, Jeff. Can sweet treats drive kids crazy? Sugar and Hyperactivity in Children 1996. *Nutrition Bytes*, 2(1). http://escholarship.org/uc/item/5pr0m8mc

Gibson, C. J. and Richard J. Wurtman. Physiological control of brain epinephrine synthesis by brain tyrosine concentration. *Life Science*, 22(1) (1978):399–1406.

McAulay, Vincent, Ian J. Deary, Stewart C. Ferguson and Brien M. Frier. Acute hypoglycemia in humans causes attentional dysfunction while nonverbal intelligence is preserved. *Diabetes Care* 24(10) (2001):1745–1750.

McAulay, Vincent, Jonathan R. Seckl and Mark W. Strachan. Recurrent dizzy spells: All in the head! *Annals of Clinical Biochemistry* 42(Part I) (2005):78–79.

Rogers, Peter J. and Helen M. Lloyd. Nutrition and mental performance. *Proceedings of the Nutrition Society, Institute of Food Research in Great Britain* 53 (1994):443–456.

Rosenthal, J. Miranda, Stephanie A. Amiel, Lidia Yáguez, et al. The effect of acute hypoglycemia on brain function and activation: A functional Magnetic Resonance Imaging study. *Diabetes* 50(7) (2001):1618–1626.

Shaywitz, B.A., C.M. Sullivan, G.M. Anderson, et al. Aspartame, behavior, and cognitive function in children with attention deficit disorder. *Pediatrics*. 93(1) (1994):70–75.

Spiers, P.A., L. Sabounjian, A. Reiner, et al. Aspartame: Neuropsychologic and neurophysiologic evaluation of acute and chronic effects. *Am J Clin Nutr* 68(3) (1998):531–537.

Galactose

Galactosemia, Isselbacher Kurt J. Galactokinase Deficiency, and other Rare Disorders of Carbohydrate Metabolism. Ch 354. *Harrison's Principles of Internal Medicine*, 13th edition, Isselbacher, et al, editors, McGraw-Hill, Inc., San Francisco, 1994. p. 2131–2132.

Shils, M, J. A. Olson and M. Shike. *Galactosemia and Galactose-Restricted Diet. Diet and Nutrition in Disease*. Eighth edition. Vo. 2. Lea and Febiger, Malvern, PA. 1994 p. 1998–1202.

Springhouse Corporation. *Illustrated Guide to Diagnostic Tests*, Springhouse Corp., Springhouse, Pennsylvania, 1994: 123–125.

Lipids

Ayton AK. Dietary polyunsaturated fatty acids and anorexia nervosa: Is there a link? *Nutr Neurosci.* 2004 Feb;7(1):1–12. Eating Disorders Unit, Huntercombe Stafford Hospital, Staffordshire, UK. PubMed 15085553. agnes.ayton@fshc.co.uk

Blasko I. Accompanying Editorial Imrich Blasco Interaction of ω-3 fatty acids with B vitamins in slowing the progression of brain atrophy: identifying the elderly at risk. *Am J Clin Nutr* 2015; 102(1):7–8; First published online June 10, 2015.

Edgar, PF, AJ Hooper, NR Poa and JR Burnett. Violent behavior associated with hypocholesterolemia due to a novel APOB gene mutation. *Mol Psychiatry* 2007; 12(3):258–263.

Favaro A, L Caregaro, L Di Pascoli, F Brambilla and P Santonastaso. Total Serum Cholesterol and Suicidality in Anorexia Nervosa. *Psychosom Med.* 2004; 66:548–552.

Fiala M, RC Holder, B Sagong, O Ross. J Sayre, V Porter and DE Bredesen. ω-3 Supplementation increases amyloid-β phagocytosis and resolvin D1 in patients with minor cognitive impairment. *The FASEB J.* 2015; 29(7):2681–2689.

Freeman MP. Omega-3 fatty acids in psychiatry: A review. *Ann Clin Psychiatry.* 2000 Sept; 12(3):159–165.

Haag M. Essential fatty acids and the brain. *Can J Psychiatry.* 2003 Apr; 48(3):195–203 https://www/scbi.nlm.nih.giv/entrez, accessed 10-23-03

Hibbeln JR. Omega-3 fatty acids and mental health. Omega-3 fatty acids: Recommendations for therapeutics and prevention symposium. 2005. New York. Medscape Oct, 2005.

Hibbeln JR, M Linnoila, JC Umhau, et al. Essential fatty acids predict metabolites of serotonin and dopamine in cerebrospinal fluid among healthy control subjects, and early- and late-onset alcoholics. *Biol Psychiatry* 1998; 44(4):235–242.

Hibbeln JR and N Salem. Dietary polyunsaturated fatty acids and depression: When cholesterol does not satisfy. *Am J Clin Nutr* 1995; 62(1):1–9, July. Nat'l Inst Alc Abuse and Alcoholism, Rockport MD.

Hibbeln R, JC Umhau, M Linnoila, et al. "A replication study of violent and nonviolent subjects: Cerebrospinal fluid metabolites of serotonin and dopamine are predicted by plasma essential fatty acids." *Biol. Psychiatry* 44 (1998):243–249.

Jerneren F and AK Elshorbagy et al. Brain atrophy in cognitively impaired elderly: the important of long chain fatty acids and B-vitamin status in a randomized controlled trial. *Am J Clin Nutr.* 2015; 102(1): 215–221.

Jones JW and M Sidwell, *Essential Fatty Acids and Treatment of Psychiatric Diseases.* Original Internist, Inc., March 2001, vol 8: 5.

Joy CD, R Mumby-Croft and LA Joy. Polyunsaturated fatty acid supplementation for schizophrenia. *Cochrane Rev Abstract* updated 4-1-03. http://www.
medscape.com article 454304.

Liu JJ, P Green, J John Mann, SI Rapoport and ME Sublette. Pathways of
polyunsaturated fatty acid utilization: Implications for brain function in
neuropsychiatric health and disease. *Brain Res.* 2015 Feb 9(1597):220–246

Martinez JM and LB Marangel. Omega-3 fatty acids: Do 'fish oils' have a ther-
apeutic role in psychiatry? *Current Psychiatry Online* Jan 2004; 3(1). http://
www.currentpsychiatry.com/2004_01/0104_omega-3_fatty_acids.asp

Oh R. Practical Applications of Fish Oil (Omega-3 Fatty Acids) in Primary
Care. *J Am Board Fam Pract.* 2005; 18(1):28–36.

Peet M, JD Laugharne, J Mellor and CN Ramchand. Essential fatty acid deficiency in erythrocyte membranes from chronic schizophrenic patients and
the clinical effects of dietary supplementation. *Prostaglandins Leukot Essen
Fatty Acids.* 1996 Aug; 55(1–2):71–75. http://www.ncbi.nlm.gov.entrez.

Peet M and DF Horrobin. A dose-ranging study of the effects of ethyl-
eicosapentaenoate in patients with ongoing depression despite apparently adequate treatment with standard drugs. *Arch Gen Psychiatry* 2002;
59:913–919.

Simopoulis AP. The importance of the ratio of omega-6/omega-3 essential
fatty acids. *Biomed Pharmacother.* 2002 Oct; 56(8):365–379. http://www.
ncbi.nih.gov.entrez

Stoll AL. Lecture to manic-depressive and depressive association of Boston.
June 9, 1999. Published in *Arch Gen Psychiatry;* May, 1999.

Tiemeier H, HR van Tuijl, A Hofman, AJ Kiliaan and MM Breteler. Plasma
fatty acid composition and depression are associated in the elderly: The
rotterdam study. *Am J Clin Nutr.* 2003 Jul; 78(1):40–46. http://www.ncbi.
nlm.nih.giv/entrez.

Vancassel S, G Durand, C Barthelemy, et al. Plasma fatty acid levels in autistic
children. *Prostaglandins Leukot Essent Fatty Acids.* 2001 Jul; 65(1):1–7. Pub
Med.

Minerals

Chromium

Davidson, Jonathan R.T., Kurian Abraham, Kathryn M. Connor, and
Malcolm N. McLeod. Effectiveness of chromium in atypical depression:
A placebo-controlled trial. *Biol Psychiatry* 53 (2003):261–264.

Docherty, John P., David Sack, Mark Roffman, Manley Finch, and James R.
Komorowski. A Double-Blind, Placebo-Controlled, Exploratory Trial
of Chromium Picolinate in Atypical Depression: Effect on Carbohydrte
Craving. *J Psychiatric Pract* 11(5) (2005):302–314.

Iodine

Hollowell JG, Staehling NW, Hannon WH, et al. Iodine nutrition in the United States. Trends and public health implications: iodine excretion data from National Health and Nutrition Examination Surveys I and III (1971–1974 and 1988–1994). *J Clin Endocrinol Metab.* Oct 1998; 83(10):3401–3408.   http://consumer.healthday.com/cognitive-health-information-26/brain-health-news-80/iodine-deficiency-common-in-pregnancy-say-pediatricians-688147.html
Raj PY Treating thyroid disorders and depression: 3 case studies. *Current Psychiatry Online.* 2013, 12(1). http://www.currentpsychiatry.com/index. php?id=22161&cHash=071010&tx_ttnews[tt_news]=177414
Sarlis N an Boaz Hirschberg. Goiter, Lithium-Induced. http://medicine. medscape.com/article/120243

Magnesium

Birmingham, C., D. P. Laird, and J. Hlynsky. Hypomagnesmia during refeeding in anorexia nervosa. *J Eating and Weight Disorders* 9(3) (2004):236–237. https://umm.edu/health/medical/altmed/supplement/magnesium
Mota de Freitas, D., M. M. Castro, and C. F. Geraldes. Is competition between Li+ and Mg2+ the underlying theme in the proposed mechanism for the pharmacological action of lithium salts in bipolar disorder? *Accounts Chem Res* 39(4) (2006):283–291.
Shaldubina, A., Z. Stahl, M. Furzpan, et al. Inositol deficiency diet and lithium effects. *Bipolar Disord* 8(2) (2006):152–159

Sodium

Bersani G, L Pesaresi, V Orlandi, S Gherardelli, P Pancheri. Atypical antipsychotics and polydipsia: a cause or a treatment? *Hum Psychopharmacol.* 2007 Mar;22(2):103–107. http://www.ncbi.nlm.nih.gov/pubmed/17335101
Hew-Butler T, M Rosner, S FowkesBodek, et al. Statement of the Third International Exercise-Associated Hyponatremia Consensus Development Conference, Carlsbad, California, 2015. *Clin J Sport Med.* July 2015 – 25(4):303–320. doi: 10.1097/JSM.0000000000000221

Selenium

Agency for Toxic Substances and Disease Registry (ATSDR). Toxicological Profile for Selenium. U.S. Department of Health and Human Services, Public Health Services, Atlanta, GA, 2003. http://www.atsdr.cdc.gov/toxprofiles/tp92. html
Benton D. Selenium intake, mood and other aspects of psychological functioning. *Nutr NeuroSci* 2002. 5 (6): 363–374.

Clark RF, E Strukle, W Saralyn, AS Manoguerra. Selenium poisoning from a nutritional supplement. *JAMA* 1996; 275:1087–1088.

FDA. Toxicity with superpotent selenium. *FDA Drug Bull* 1984;14:19.

Glover JR. Selenium and its industrial toxicology. *Ind Med Surg.* 1970;39:50–54.

Hatchcock JN and JI Rader. Food additives, contaminants and natural toxins. In *Modern Nutrition in Health and Disease.* 1593. Ed. ME Shils, JA Olson, and M Shike. Philadelphia: Lea & Febiger, 1994: 1593.

Holben DH and AM Smith. The Diverse Role of Selenium with Selenoproteins: A Review. *J Amer Diet Assoc.* 1999. 99(7):836–843. http://ods.od.nih.gov/factsheets/Selenium-HealthProfessional/ accessed 1-6-14

Kibriya MG. Changes in gene expression profiles in response to selenium supplementation among individuals with arsenic-induced pre-malignant lesions. *Toxicol Lett.* 2007; 169(2):162–176.

Longnecker MP et al. Selenium in diet, blood, and toenails in relation to human health in seleniferous area. *Am J Clin Nutr* 1993; 53:1288–1294.

Lowry F. Dietary supplement causes widespread selenium poisoning. Sept 18, 2010. http://www.medscape.com/viewarticle/716598

Milman N, Laursen J, Byg KE et al. Elements in autopsy liver tissue samples from Greenlandic Inuit and Danes. V. Selenium measured by X-ray fluorescence spectrometry. *J Trace Elem Med Biol* 2004;17:301–306.

Nuttall KL Evaluating selenium poisoning. *Ann Clin Lab Sci.* 2006; 36(4):409–420.

Pasco JA et al. Dietary selenium and major depression: A nested case-control study. *Complement Ther Med.* 2012; 20(3):119–123. doi: 10.1016/j.ctim.2011.12.008. Epub 2012 Jan 30.

Ravaglia G, P Forti, F Maioli et al. Bastagli. Effect of micronutrient status on natural killer cell immune function in healthy freeliving subjects aged ≥90 y. *Amer J Clin Nutr* 2000; 71(2):590–598. PMID 10648276.

Rayman M et al. Impact of selenium on mood and quality of life. *Biol Psychiatry* 2006. 15(2): 147–154.

Sauberlich HE *Assessment of Nutritional Status.* 1999 CRC Press. New York.

Vaddadi KS, E Soosai and G Vaddadi. Low blood selenium concentrations in schizophrenic patients on clozapine. *Br J Clin Pharmacol.* 2003 March; 55(3): 307–309. doi: 10.1046/j.1365-2125.2003.01773.x

Yanik M. et al. Plasma manganese, selenium, zinc, copper, and iron concentrations in patients with schizophrenia. *Biol Trace Elem Res.* 2004 May; 98(2):109–117. Virginia Dept. of Health http://www.vdh.state.va.us/news/Alerts/Selenium/SeleniumFAQs.htm

Zinc

Beard, John. Nutrient status and central nervous system function. In *Present Knowledge in Nutrition*, 7th ed., Edited by Ekhard E. Ziegler and L. J. Filer, Jr., 612–622. Washington, DC: International Life Sciences Institute (ILSI) Press, 1996.

Bogden, John D. and Raymond A. Troiano. Plasma calcium, copper, magnesium, and zinc concentrations in patients with the alcohol withdrawal syndrome. *Clinical Chemistry* 24 (9) (1978):1553–1556.

Levenson, Cathy W. Zinc: The new antidepressant? *Nutrition Reviews* 64(1) (2006): 39–42.

Minerals – Toxic

Araujo, J, AP Beelen, LD Lewis, et al. Lead poisoning associated with Ayurvedic medications- five states 2000–2003. *MMWR*. 2004; 53(26):582–584.

Hightower, JM and D Moore. Mercury levels in high-end consumers of fish. *Environ Hlth Persp* 2003; 11(4):604–608. http://www.nrdc.org/health/effects/mercury/guide.asp

Lynch, R, B Elledge, and C Peters. An assessment of lead leachability from lead-glazed ceramic cooking vessels. *J Environ Health*. 2008; 70(9):36–40, 53. http://www.ncbi.nlm.nih.gov/pubmed/18517152

Saper, RB, SN Kales, J Paquin, et al. Heavy metal content of ayurvedic herbal medicine products. *JAMA*. 2004 Dec 15;292(23):2868–2873.

Vitamins

Vitamin A

http://ods.od.nih.gov/factsheets/VitaminA-HealthProfessional/

Vitamin B-1 – Thiamin

Accetta SG, Abeche AM, Buchabqui JA, Hammes L, Pratti R, Afler T, Capp E. Memory loss and ataxia after hyperemesis gravidarum: a case of Wernicke-Korsakoff syndrome. *Eur J Obstet Gynecol Reprod Biol*. 2002 Apr 10;102(1):100.

Angstadt JD, Bodziner RA. Peripheral polyneuropathy from thiamine deficiency following laparoscopic Roux-en-Y gastric bypass. *Obes Surg*. 2005 Jun–Jul;15(6):890–892.

Benton D, Griffiths R, Haller J. Thiamine supplementation improves mood and cognitive functioning. *Psychopharmacology* 1997;129:66–71.

Cirignotta F, Manconi M, Mondini S, Buzzi G, Ambrosetto P. Wernicke-korsakoff encephalopathy and polyneuropathy after gastroplasty for morbid obesity: report of a case. *Arch Neurol*. 2000 Sep;57(9):1356–1359.

Galvin R, Brathen G, Ivashynka A, et al. EFNS guidelines for diagnosis, therapy and prevention of Wernicke encephalopathy. *Eur J Neurol*. 2010;17(12):1408–1418. http://www.ncbi.nlm.nih.gov/pubmed/20642790

Gilchrist, de la Monte S. Alcohol-related peripheral neuropathy: Nutritional, toxic, or both? *Muscle Nerv* 2011;43(3):309–316.

Koike H, Lijima M, Sugiura M, et al. Alcoholic neuropathy is clinico-pathologically distinct from thiamin-distinct neuropathy. *Ann Neurol* 2003;54(1):19–29.

Lin I, Lin YL. Peripheral polyneuropathy after bariatric surgery for morbid obesity. *J Family Community Med.* 2011 Sep-Dec;18(3): 162–164. (text + images)   http://www.ncbi.nlm.nih.gov/pmc/articles/PMC3237207/? report=printable

Lingford-Hughes AR, Welch S, Nutt DJ. Evidence-based guidelines for the pharmacological management of substance misuse, addition and comorbidity: Recommendations for the British Association for Psychopharamacology. *J Psychopharmacology* 2004;18(3):293–335.

Liu IC, Blacker DL, Xu RP, et al. Genetic and environmental contributions to the development of alcohol dependence in male twins. *Arch Gen Psychiatry* 2004;6:897– 903.

Loh Y, Watson WD, Verma A, Chang ST, Stocker DF, Labutta RJ. Acute Wernicke's encephalopathy following bariatric surgery: clinical course and MRI correlation. *Obes Surg.* 2004 Jan; 14(1): 129–132.

Sgouros, MB, et al., Evaluation of a clinical screen instrument to identify states of thiamine deficiency in inpatients with severe alcohol dependence syndrome. *Alcohol and Alcoholism,* 2004 May; 39(3):227–232. https://doi. org/10.1093/alcalc/agh051.

Victor M, Martin JB. Nutritional and Metabolic Diseases of the Nervous System. Ch. 377: 2329–2333. *Harrison's Principles of Internal Medicine* KJ Illelbacher, E Braunwald, JD Wilson, JM Martin, AS Fauci and DL Kaspar Ed. 13th edition. McGraw-Hill, Inc., San Francisco 1994.

Wilson JD. Vitamin Deficiency and Excess. *Harrison's Principles of Internal Medicine.* Chapter 77: 472–480. Illelbacher KJ, Braunwald E, Wilson JD, Martin JM, Fauci AS and Kaspar DL, Ed. 13th edition, McGraw-Hill, Inc., San Francisco 1994.

Wurtman R, Wurtman J Ed: *Nutrition and the Brain.* Raven Press NY 1980, p.142. found in: Wolstenholme, GEW and O'Connor, Meds: Thiamin Deficiency: Biochemical Lesions and their Clinical Significance. CIBA Foundation Study Group. No. 28. Little, Brown, and Co. Boston, 1967.

Vitamin B-2 – Riboflavin

Haller, Jurg. Vitamins and Brain Function. In *Nutritional Neuroscience.* Edited by Harris R. Lieberman, Robin B. Kanarek, and Chandan Prasad, 229. Boca Raton, FL: CRC Press: Taylor & Francis Group, 2005.

Vitamin B-3 – Niacin

Bourgeois, Christelle, Daniel Cervantes-Laurean, and Joel Moss. Niacin. In *Modern Nutrition in Health & Disease.* Edited by Maurice E. Shils, Moshe Shike, A. Catherine Ross, Benjamin Caballero, and Robert J. Cousins 445–450. Wilkins, New York: Lippincott Williams & Wilkins, 2005.

Mark, A. Oldham and Ana Ivkovic. Pellagrous encephalopathy presenting as alcohol withdrawal delirium: A case series and literature review. *Addict Sci Clin Pract.* 2012; 7(1):12. Published online 2012 Jul 6. doi: 10.1186/1940-0640-7-12

Morris, Martha Clare. Dietary niacin and risk of incident Alzheimer's Disease and of cognitive decline. *Journal of Neurology, Neurosurgery and Psychiatry* 75 (2004):1093–1099.

Victor, Maurice and Joseph B. Martin. Nutritional and Metabolic Diseases of the Nervous System. In *Harrison's Principles of Internal Medicine.* Edited by Kurt J. Isselbacher, Eugene Braunwald, Jean D. Wilson, Joseph B. Martin, Anthony S. Fauci, and Dennis L Kaspar. p. 2329–2333. San Francisco: McGraw-Hill, Inc., 1994.

Wilson, Jean D. Vitamin deficiency and excess. In *Harrison's Principles of Internal Medicine.* Edited by Kurt J. Isselbacher, Eugene Braunwald, Jean D. Wilson, Joseph B. Martin, Anthony S. Fauci, and Dennis L. Kasper. 472–480. San Francisco: McGraw-Hill Inc., 1994.

Vitamin B-6 – Pyridoxine

Hunt, Sara M and James L Groff. *Advanced Nutrition and Human Metabolism.* 212–280. Los Angeles: West Publishing Company, 1990.

Vitamin B12 – Cobalamin

Carmel, Ralph. Prevalence of undiagnosed pernicious anemia in the elderly. *Arch Inter Med* 156(10) (1996):1097–1100.

Eastley, Rebecca, Gordon K. Wilcock, and Ramola S. Bucks. Vitamin B-12 deficiency in dementia and cognitive impairment: The effects of treatment on neuropsychological function *Int'l J Geriatric Psychiatry* 15(3) (2000):226–233.

Garcia, Angela, Alicia Paris-Pombo, Lisa Evans, Andrew Day, and Morris Freedman. Is low- dose oral cobalamin enough to normalize cobalamin function in older people? *J Amer Geriatric Soc* 50(8) (2002):1,401–1,404.

Hunt, Sara M and James L. Groff. *Advanced Nutrition and Human Metabolism.* 212–280. Los Angeles: West Publishing Company, 1990.

Lam, J.R., J. L. Schneider, W. Zhao, D. A. Corley. Proton pump inhibitor and histamine 2 receptor antagonist use and vitamin B12 deficiency. *JAMA.* 310(22) (2013 Dec 11):2435–2442. doi: 10.1001/jama.2013.280490.

Louwman, Marieka W.J., Marijke van Dusseldorp, Fons J.R. van de Vijver, Chris MG Thomas, Jorn Schneede, Per M. Ueland, Helga Refsum, and Wija A van Staveren. Signs of Impaired function in adolescents with marginal cobalamin status. *American Journal Clinical Nutrition* 2000; 72(3):762–769. http://ods.od.nih.gov/factsheets/VitaminB12-HealthProfessional/

Zucker, D. K., R. L. Livingston, R. Nakra, and P. J. Clayton. B12 deficiency and psychiatric disorders: Case report and literature review. *Biological Psychiatry* 16(2) (1981):197–205.

Biotin

Jones, Walretta O. and Bernard D. Nidus. Biotin and hiccups in chronic dialy-sis patients. *Journal of Renal Nutrition* 1(2) (1991'):80–83.
Scheinfeld, Noah S. and Stephanie Beth Freilich. Biotin Deficiency. http://www.emedicine.com/ped
Staggs, C.G., W.M. Sealey, B. J. McCabe, A. M. Teague and D. M. Mock. Determination of the biotin content of select foods using accurate and sensitive HPLC/avidin binding. *J Food Compost Anal.* 17(6) (2004 Dec):767–776.

Vitamin C – Ascorbic acid

Kinsman, Robert A. and James Hood. Some behavioral effects of ascorbic acid deficiency. *Amer J Clinl Nutr* 24 (1971):455–464.
Roy-Byrne, Peter, D. A. Gorelick, and Stephen R. Marder. Unusual dietary habits in a patient with schizotypal personality disorder: Interaction of nutritional status and psychopathology. *J Psychiatric Treatment and Eval* 5 (1983):67–69.
Walter, Joseph F. Scurvy resulting from a self-imposed diet. *Western J Med* 130(2) (1979):177– 179.
Wilson, Jean D. Vitamin deficiency and excess. In *Harrison's Principles of Internal Medicine.* Edited by Kurt J. Isselbacher, Eugene Braunwald, Jean D. Wilson, Joseph B. Martin, Anthony S. Fauci, and Dennis L. Kasper. 472–480. San Francisco: McGraw-Hill Inc., 1994.

Vitamin D

Creighton University. Recommendation for vitamin D intake was miscalcu-lated, is far too low, experts say. *Science Daily.* 17 March 2015.
Ganji, V., C. Milone, M. M. Cody et al. Serum vitamin D concentrations are related to depression in young adult U.S. population: The Third National Health and Nutrition Examination survey. *Int'l Arch Med* 2010; 3:29. http://www.intarchmed.com/content/3/1/29
Patrick, Rhonda P. and Bruce Ames. Vitamin D and the omega-3 fatty acids control serotonin synthesis and action, part 2: relevance for ADHD, bipolar disorder, schizophrenia, and impulsive behavior. *FASEB J.* 2015; 29(6):2207–2222.
Sepehrmanesh, Azhra, Fariba Kolahdooz, Fatemeh Abedi, et al. Vitamin D Supplementation Affects the Beck Depression Inventory, Insulin Resistance, and Biomarkers of Oxidative Stress in Patients with Major Depressive Disorder: A Randomized, Controlled Clinical Trial. 2016; *Am J Nutr* 146(2):243–248. doi: 10.3945/jn.115.218883.
Simopoulis, A.P. The importance of the ratio of omega-6/omega-3 essential fatty acids. *Biomed Pharmacother.* 2002; 56:365–379.

Valipour, G., Saneei, P., Esmaillzadeh, A. Serum vitamin D levels in relation to schizophrenia: a systematic review and meta-analysis of observational studies. *J Clin Endocrinol Metab*. 2014 Oct;99(10):3863–3872.

Veugelers, Paul and John Ekwaru. A Statistical Error in the Estimation of the Recommended Dietary Allowance for Vitamin D. *Nutrients*, 2014; 6 (10): 4472. doi: 10.3390/nu6104472

Vitamin D Council http://www.vitaminDCouncil.org

Vitamin E

Owen, AJ, MJ Batterham, YC Probst, et al. Low plasma vitamin E levels in major depression: Diet or disease? *Eur J Clin Nutr* 2005; 59(2):304–306.

Folacin

Bottiglieri, Teodoro, Malcom Laundry, Richard Crellin, Brian K. Toone, Michail W.P. Carney, and Edward H. Reynolds. Homocysteine, folate, methylation and monoamine metabolism in depression. *J Neur, Neurosurg Psychiatry* 69 (2000):228–232.

Goff, Donald C., Teodoro Bottiglieri, Erland Arning, Vivian Shih, Oliver Freudenreich, Eden Evins, David C. Henderson, Lee Baer, and Joseph Coyle. Folate, homocysteine, and negative symptoms in schizophrenia. *Amer J Psychiatry* 161 (2004):1705–1708. http://lpi.oregonstate.edu/mic/vitamins/folate Reviewed in December 2014

James, S. Jill, Paul Cutler, Stepan Melnyk, Stefanie Jernigan, Laurette Janak, David W Gaylor and James A Neubrander. Metabolic biomarkers of increased oxidative stress and impaired methylation capacity in children with autism. *Amer J Clin Nutr* 80(6) (2004):1611–1617.

Office of Dietary Supplements. http://ods.od.nih.gov/factsheets/Folate-Health Professional/

Smith, A.D., Y. I. Kim, H. Refsum Is folic acid good for everyone? *Am J Clin Nutr*. 2008 Mar;87(3):517–533.

Other

Chawla, Jasvinder, and Nicholas Lorenzo. Nutritional Neuropathy. Updated January 2013. http://emedicine.medscape.com/article/1171558-overview, https://sparck.nationalacademies.org/vivisimo/cgi-bin/query-meta?

Quirk, Shae E, Lana J. Williams, Adrienne O'Neil, et al. The association between diet quality, dietary patterns and depression in adults: a systematic review. *BMC Psychiatry* 2013, 13:175. http://www.biomedcentral.com/1471-244X/13/175

Caffeine

European Food Safety authority report on Caffeine; May 27, 2015. http:// www.efsa.europa.eu/en/efsajournal/pub/4102.htm

Supplements

Sarris J, D Mischoulon, I Schweitzer. Adjunctive nutraceuticals with standard pharmacotherapies in bipolar disorder: a systematic review of clinical trials. *Bipolar Disord.* 2011 Aug– Sep;13(5–6):454–465. doi: 10.1111/ j.1399-5618.2011.00945.x.

Brain Plasticity

Goh, Joshua O. and Denise C. Park. Neuroplasticity and cognitive aging: The scaffolding theory of aging and cognition. *Restorative Neurology and Neuroscience,* 2009; 27(5): 391–403.

Research

de Jager CA, L Dye , EA de Bruin, L Butler, J Fletcher, DJ Lamport, ME Latulippe, JP Spencer, K Wesnes. Cognitive function: criteria for validation and selection of cognitive tests for investigating the effects of foods and nutrients. *Nutr Rev.* 2014 Feb 22. doi: 10.1111/nure.12094.

Leyse-Wallace, Ruth. *Linking Nutrition to Mental Health.* iUniverse, Inc., Lincoln, NE. 2008.

Leyse-Wallace, Ruth. *Nutrition and Mental Health.* Taylor and Francis, Inc., CRC Press. Boca Raton, FL. 2013.

Index

Pages in *italics* refer to figures and pages in **bold** refer to tables.